Mattia Bertschi

Electrical and mechanical activity of the intestine

Mattia Bertschi

Electrical and mechanical activity of the intestine

Experimental assessment and biophysical modeling

Südwestdeutscher Verlag für Hochschulschriften

Impressum/Imprint (nur für Deutschland/ only for Germany)
Bibliografische Information der Deutschen Nationalbibliothek: Die Deutsche Nationalbibliothek
verzeichnet diese Publikation in der Deutschen Nationalbibliografie; detaillierte bibliografische
Daten sind im Internet über http://dnb.d-nb.de abrufbar.
Alle in diesem Buch genannten Marken und Produktnamen unterliegen warenzeichen-, marken-
oder patentrechtlichem Schutz bzw. sind Warenzeichen oder eingetragene Warenzeichen der
jeweiligen Inhaber. Die Wiedergabe von Marken, Produktnamen, Gebrauchsnamen,
Handelsnamen, Warenbezeichnungen u.s.w. in diesem Werk berechtigt auch ohne besondere
Kennzeichnung nicht zu der Annahme, dass solche Namen im Sinne der Warenzeichen- und
Markenschutzgesetzgebung als frei zu betrachten wären und daher von jedermann benutzt
werden dürften.

Verlag: Südwestdeutscher Verlag für Hochschulschriften Aktiengesellschaft & Co. KG
Dudweiler Landstr. 99, 66123 Saarbrücken, Deutschland
Telefon +49 681 37 20 271-1, Telefax +49 681 37 20 271-0, Email: info@svh-verlag.de
Zugl.: Lausanne, Swiss Federal Institute of Technology, Ph.D., 2005

Herstellung in Deutschland:
Schaltungsdienst Lange o.H.G., Berlin
Books on Demand GmbH, Norderstedt
Reha GmbH, Saarbrücken
Amazon Distribution GmbH, Leipzig
ISBN: 978-3-8381-0505-5

Imprint (only for USA, GB)
Bibliographic information published by the Deutsche Nationalbibliothek: The Deutsche
Nationalbibliothek lists this publication in the Deutsche Nationalbibliografie; detailed
bibliographic data are available in the Internet at http://dnb.d-nb.de.
Any brand names and product names mentioned in this book are subject to trademark, brand or
patent protection and are trademarks or registered trademarks of their respective holders. The
use of brand names, product names, common names, trade names, product descriptions etc.
even without a particular marking in this works is in no way to be construed to mean that such
names may be regarded as unrestricted in respect of trademark and brand protection legislation
and could thus be used by anyone.

Publisher:
Südwestdeutscher Verlag für Hochschulschriften Aktiengesellschaft & Co. KG
Dudweiler Landstr. 99, 66123 Saarbrücken, Germany
Phone +49 681 37 20 271-1, Fax +49 681 37 20 271-0, Email: info@svh-verlag.de

Copyright © 2009 by the author and Südwestdeutscher Verlag für Hochschulschriften
Aktiengesellschaft & Co. KG and licensors
All rights reserved. Saarbrücken 2009

Printed in the U.S.A.
Printed in the U.K. by (see last page)
ISBN: 978-3-8381-0505-5

*To Arianna and
my little Amélie*

Acknowledgments

A thesis is an original work of research in a specific field, written by a single person, but only possible with team work. I thus want to thank sincerely all the people that, in a way or another, participated to the success of this project:

- First of all I would like to thank my thesis director, Prof. Murat Kunt (Signal Processing Institute, Swiss Federal Institute of Technology, Lausanne), for accepting me in his laboratory that offers an international and multicultural environment, optimal conditions for the research and the innovation.

- A special thanks to Dr. Jean–Marc Vesin, who has not only been my thesis director, but also a precious friend. During the last four years, his incredible scientific knowledge has represented a certitude through the many doubts in the research and his sane curiosity as a powerful motivator during the difficult moments.

- In the framework of the partnership with Medtronic Europe (Tolochenaz), I had the opportunity and the luck to meet Dr. Nathalie Virag. Her sincerity, diplomacy, sense of organization, and constructive criticism have been of primary importance for the resolution of this work.

- I would like to thank Prof. Pavel Kučera (Institute of Physiology, University of Lausanne) and his assistant Dr. Vincent Schlageter, for their guidance their encouragement, and for their relevant comments and suggestions during our fruitful collaboration.

- My thesis work would not have been possible without the contribution of the ColoStim team, a multidisciplinary group born from the collaboration between the EPFL, the UNIL, the CHUV, and Medtronic Europe.

- Thank you to all the people of the Signal Processing Institute and particularly to my collegues and friends of the ELE–227 room for the wonderful atmosphere: Dr. Vincent Jacquemet, Zenichi Ihara, Lam Dang, Mathieu Lemay, Philippe Jost, and Yann Prudat. I am particularly grateful to Dr. Olivier Blanc for introducing me to the biophysical models and for the very interesting discussions. I would also like to thank Elisa Drelie who had the difficult role of sharing the office with me and who had to suffer my changes of mood during these last six months.

- I would like to express my sincere gratitude to the LTS staff members for their help at many stages of my work: Marianne Marion, Fabienne Vionnet, and Gilles Auric.

- The friendship is important, and important are: Teo the Captain, Gio the Doctor, Luciana Jones and his extraordinary adventures, Scola because precision comes first of all, and Mao.

- My gratitude goes to my parents who have always encouraged me during my whole study and who made everything possible to realize my dreams, and to Arianna who shared with me the good and the bad times during... many years and to have given me the most wonderful of the gift, the little Amélie.

Mattia Bertschi
Lausanne, 18th May 2005

Abstract

The intestine is today still an organ whose mechanisms of functions are not fully understood. The difficulty in its understanding comes from the fact that it is regulated by a complex network of signals of different natures. Granting that a multitude of experimental techniques have brought precious information on the electrophysiological processes associated to the intrinsic activity of the intestine, there is still shadow covering over the boundary of knowledge.

With a view to sharpening the study of the complex spatio–temporal dynamics of the intrinsic electrical and mechanical activity of the intestine, three different numerical models of intestinal tissue were developed. Of particular interest are the networks of coupled oscillators, exhibiting a mass behaviour with marked analogy to the intestinal tissue. The conception of these models were conducted by the handling of effective numerical methods and evaluation of numerical instability and errors. A simplified geometrical model of human colon was also developed, so as to allow a realistic visualization of the simulated results.

Experimental measures *in vitro* and *in vivo* of an animal model, namely the pig, endorsed us the recordings of intrinsic electrical and mechanical activity of different intestinal segments. The recordings of the mechanical activity were obtained by three different tools: visual inspection, manometry, and Magnet Tracking. The first method implies the use of image processing for the evaluation of intestinal movement. The second one provides local pressure measurements. Only the third method enables non–invasive measurements along the digestive tube. In order to automate the extraction of valuable information contained in the signals recorded by the last method, an approach was proposed based on morphological filtering.

These three methods of measure were also used in a parallel study to evaluate the electrical stimulation of the colon, with a goal to find an alternative solution to classical treatments of chronic constipation. We showed that it is possible to induce local contraction in the caecum with electrical stimulation generated by an implantable device.

Computer models guarantee at any time the realization of reproducible simulations of technically difficult, and even impossible, experiments. They compel, nonetheless, a simplification of the physiological reality. So as to validate the computer models, the result of simulations must be treated by particular methods. Electrograms and signals representing the mass movement were computed and faced to observations and real measurements. Simulated electrograms revealed a similarity to those recorded on animal intestines. Numerous resemblances are likewise observable between simulated signals of mass movement and the recordings of Magnet Tracking.

Contents

Acknowledgments	i
Abstract — Version Abrégée	iii
1 Introduction	**1**
1.1 Motivations and Problem Statement	1
1.2 Organization of the Dissertation	3
1.3 Original Contributions	3
I General Principles of Gastro–Intestinal Physiology	**5**
2 The Gastro–Intestinal Smooth–Muscle Cell	**7**
2.1 Cell Membrane	8
2.1.1 Passive Transport	9
2.1.2 Active Transport	9
2.1.3 Measuring Ionic Channel Current	10
2.2 Membrane Potential	11
2.2.1 The Nernst Equation	11
2.2.2 Resting Membrane Potential and Action Potentials	12
2.2.3 Measuring Membrane Potential	12
2.3 Slow Waves and Spike Potentials	13
2.4 Conclusion	14
3 The Human Digestive System	**15**
3.1 Structure	15
3.2 Functions	18
3.2.1 Motility	18
3.2.2 Secretion	20
3.2.3 Digestion	22
3.2.4 Absorption	23
3.2.5 Immuno–Defense	25

	3.3	Control	25
		3.3.1 The Enteric Nervous System	25
		3.3.2 The Enteric Endocrine System	26
	3.4	Conclusion	27

II Biophysical Modeling   29

4 Modeling Intestinal Electrophysiology   31
- 4.1 Bardakjian Model 32
  - 4.1.1 The Generic Equation of the Isolated Oscillator 33
  - 4.1.2 The Generic Equation for a Population of Bidirectionally Coupled Oscillators 36
- 4.2 Miftakhov Model 39
- 4.3 Aliev Model 46
- 4.4 Conclusion 52

5 Numerical Methods   53
- 5.1 Time Integration 54
- 5.2 Spatial Discretization 57
- 5.3 Numerical Instabilities and Errors 58
- 5.4 Conclusion 63

6 Relating Simulations to Experimental Signals   65
- 6.1 Space–Time Maps 65
- 6.2 Synchronization and Space Organization 67
  - 6.2.1 Cross–Covariance Analysis 68
  - 6.2.2 Cluster Evolution Analysis 69
- 6.3 Electrograms 70
- 6.4 Mass Displacement 73
- 6.5 Three–Dimensional Visualization 75
- 6.6 Conclusion 77

III Validation and Applications   79

7 Intestinal Electrical Activity   81
- 7.1 *In Vivo* Acute Experiments 81
  - 7.1.1 Preparation of Animals 81
  - 7.1.2 Description of the Equipment 82
  - 7.1.3 Results 82
- 7.2 Simulation of the Electrical Activity Using Computer Models 84

		7.3	Conclusion	87
8	**Intestinal Mechanical Activity**			**89**
	8.1		*In Vitro* Experiments	89
		8.1.1	Description of the Equipment	89
		8.1.2	Results	90
	8.2		*In Vivo* Experiments	91
		8.2.1	Description of the Equipment	91
		8.2.2	Description of the Data Processing Methods	92
		8.2.3	Results	93
	8.3		Simulation of the Mechanical Activity Using Computer Models	98
	8.4		Conclusion	102
9	**Electrical Stimulation**			**103**
	9.1		CTI Project Presentation	105
	9.2		*In Vitro* Experiments	106
		9.2.1	Description of the Equipment	106
		9.2.2	Results	106
	9.3		*In Vivo* Acute Experiments	107
		9.3.1	Preparation of Animals	107
		9.3.2	Description of the Equipment	108
		9.3.3	Description of the Data Processing Methods	110
		9.3.4	Results	112
	9.4		*In Vivo* Chronic Experiments	118
		9.4.1	Preparation of Animals	118
		9.4.2	Description of the Equipment	118
		9.4.3	Results	118
	9.5		Conclusion	120
10	**Conclusion**			**121**
Biblio				**124**

DISSERTATION

Chapter 1

Introduction

1.1 Motivations and Problem Statement

Clinical Considerations

Digestive diseases represent probably the second major medical problem after cardiovascular diseases. Motility is one of the principal functions of the digestive system and it is clear that without good motility not only our digestion but also our psychical condition will be altered. The motility disorders are of organic or functional origin and display numerous forms such as dysphagia, gastroparesis, ileus, irritable bowel syndrome, diarrhea, and chronic constipation.

Slow transit constipation is the most major symptom among digestive disorders. Chronic constipation is caused by a multitude of pathologies, e.g. diet, anatomic malformation, infection, or tumour. It is expressed by a decrease in the frequency of fecal matter and/or evacuation difficulties. In most cases, it is provoked by a slow transit of waste in the last part of the intestine, i.e., in the colon.

The chronic constipation represents also a major medical problem, by the number of patients concerned (10–20% of the Swiss population) as well as the diagnostic and therapeutic difficulties characterizing it. Conventional treatments consist of conservative approaches, such as modification of diet, use of laxatives, and in the worst cases, surgical interventions, generally as partial colonic resection. They are, however, of limited efficacy and the last method is radical and only empirical. Unfortunately, no alternatives to the therapies mentioned above are available for the time being. The possibility to better understand the behaviour of the intestine would allow the physicians to offer more focalized and less empirical treatments.

In Vivo **Experimentation**

The diagnosis of digestive pathologies is based on several recognized exploratory approaches. As there is no *in vivo* method to measure the neuronal or motor function directly, the clinical approaches are based on extracellular electromyography, intraluminal manometry, and imaging. Each of these techniques has its own indications, advantages and disadvantages. Indeed, luminal content displacement along the gastrointestinal tract remains difficult to study.

In vivo experimentation is part of the tools chosen to investigate the intrinsic electrical and mechanical activity of the intestine. Experiments were performed at the Animal Experimentation Center at the Lausanne University Hospital. The acute and chronic protocols had been approved by the local committee on animal care (Office Vétérinaire Cantonal de Lausanne). All animals were treated in compliance with the Principle of laboratory animal care formulated by the National Society for Medical Research and the Guide for the care and use of laboratory animals prepared by the National Academy of Sciences and published by the National Institute of Health.

Integrated Research

In order to be effective in the development of innovative therapeutic strategies, it is needed to better integrate the knowledge and progress from both clinical research and basic sciences. Computer models of the intestine are an example of such integrated research. Rooted in mathematics, biophysics, engineering, electrophysiology and gastroenterology, these models are constructed from data collected at several spatial scales (ionic channels, cell, tissue, organ) and help bridge these scales. For instance, the effect of a modification at a cellular level, such as pharmacological intervention, on the motility of the organ can be investigated.

Despite their inherent limitations and trade–offs between accuracy of the representation of electrophysiological details and computational requirements, computer models are believed to provide a framework to test hypotheses and suggest new experimental or clinical studies. Numerical simulations, or *in silico* experiments, would then become a complementary approach to the *in vitro* and *in vivo* experimentations.

***In Silico* Experimentation**

The idea to use computer models to reproduce the behaviour of the intestinal tissue is not new, but the ever–increasing computer power made it possible to simulate models of full organs with realistic size and a detailed description of the cellular electrophysiological properties. Moreover, thanks to the development of microscopical methods allowing detailed measurements of the cell membrane properties, such as the patch–clamp, it is now possible to design more realistic models.

When compared to real experiments, computer models present several advantages, but also some drawbacks. The reproducibility of experiments and results is obvious with computer simulations, while very difficult in real experimentation. The same considerations are valid for the access of data. Only a small set of parameters is accessible in electrophysiological experiments and some regions may be inaccessible, while computer models permit a complete control of every variable, at any time, and at any location. Among the advantages of the computer simulations, one can enumerate the simplicity and the reliability of the tissue preparation and the alteration of their properties. There are of course several drawbacks linked to the use of computer models, and the main one is certainly the relevance to the real phenomenon, for which the validation step is crucial.

Objectives

The objectives of the present study are:

- to record and analyse the electromechanical intrinsic activity of the intestine;

- to develop a computer model of the intestine with which the mechanisms of spatio–temporal organization can be studied;
- to compute electrograms and motility measurements from computer simulations to compare with those of real experiments;
- to study the efficacy of *in vivo* electrostimulation for the slow transit constipation treatment.

1.2 Organization of the Dissertation

This dissertation is organized in three parts:

I. General Principles of Gastro–Intestinal Physiology: the physiological, anatomical, and functional aspects of the gastro–intestinal system are briefly presented, so as to acquire the basic but essential elements used along this work. An introduction of the phenomena involved in the concept of the membrane potential is proposed in chapter 2, and chapter 3 is dedicated to more general aspects of the entire gastro–intestinal system.

II. Computer Models: a description of three major electrophysiological models of the bowel is detailed in chapter 4. The spatial and temporal discretization schemes that solve the reaction–diffusion system representing the electrical propagation in the intestine are presented in chapter 5, as well as the numerical instabilities and errors implied in the discretization step. The approaches for comparing simulated data and recorded real signals are detailed in chapter 6, using existing measurement systems of the electromechanical activity.

III. Validation and Applications: A comparison between simulation results and real measurements are presented in chapter 7 for the electrical activity and in chapter 8 for the mechanical activity. *In vitro* and/or *in vivo* experiments have been used to collect the real electromechanical signals. Chapter 9 describes the results obtained during the project at the origin of this thesis. The goal of this project was to study the feasibility of an electrostimulation device to induce intestine muscular contractions.

1.3 Original Contributions

Despite its crucial importance for the well–being of the human population, the dynamics of the GIS still raises many questions and the literature on this topic is quite scarce in the biomedical engineering journals. This thesis contributes to elucidate some of these questions and paves the way to further developments.

The main contributions of this work are:

1. A detailed presentation of three electrophysiological models of the bowel. To the best of our knowledge, it is the first time that a systematic comparison of the existing computer models is performed.

2. The introduction of a colonic tissue computer–model, consisting of modified version of an existing small–intestine model. As such, it constitutes the first elaborate computer model

of the entire colon able to reproduce the main features of the electromechanical activity of this organ.

3. An evaluation of numerical instabilities and errors in the Aliev model, that is the basis of the model developed here. This innovative analysis had never been carried out on such models.

4. The development of a simple and robust tool for the evaluation of the spatio–temporal organization of intestinal activity. This approach allowed us to highlight the degrees of organization of the various segments of the GIS in the computer model, and to provide a reasonable explanation to experimental data. This tool can be easily adapted for the study of dynamics of other physiological systems.

5. A simulation of electrograms and a model of mass displacement for the validation of *in silico* results. To the best of our knowledge, it is the first time that such electrograms have been simulated for the small intestine. Also, mass displacement simulation yields results strikingly similar to experimental data, and may lead to a better understanding of the physiological recordings.

6. A 3D geometrical model for the visualization of simulated electromechanical activity.

7. An assessment of *in vitro* and *in vivo* electromechanical activity. Du to the experimental difficulties in recording those activities, this study is of significant importance for researchers in this field.

8. An investigation of the electrostimulation of the intestine for provoking muscular contractions. This investigation, that has allowed us to identify suitable stimulation parameter ranges, constitutes a basis for further works in this domain of significant clinical importance.

Part I

General Principles of Gastro–Intestinal Physiology

This first part of the thesis is dedicated to the physiological, anatomical, and functional aspects of the human gastro–intestinal system. This part has not the pretension to cover all the particularities of the gastro–intestinal system, which is a large and very complex one. The intention is to introduce the subject, in order to provide the concepts necessary to the comprehension of the arguments proposed along the dissertation. The exploration starts in the first chapter with the electrophysiological properties of the smooth muscle cell. A brief introduction to the phenomena involved in the generation of membrane potentials is proposed, so as to acquire the basic but essential elements used further. Zooming out from cell to organs, the second and last chapter of this part shows the precise organization existing among different actors of the gastro–intestinal system and sustaining the special needs of the human body.

Chapter 2

The Gastro–Intestinal Smooth–Muscle Cell

Muscle is a body tissue consisting of long cells, known as muscle fibres, that contract when stimulated and produces motion. The primary purpose of a muscle tissue is for the physical displacement of the body parts, or substances inside the body. There are three general types of muscle: skeletal muscle, cardiac muscle, and smooth muscle. Cardiac muscle, found in the heart, and skeletal muscle, attached to the skeleton and allowing active body movements, are both usually identified as striated muscles, in contraposition to smooth muscle that is named because no striations are visible in it.

Smooth muscle is an involuntary muscle, i.e. it contracts without conscious control, and can be found in many organs of the body, such as blood vessels, the gastro–intestinal tract, the bladder, or the uterus. Composed of far small fibres, usually 2 to $5\mu m$ in diameter and only 20 to $500\mu m$ in length, the smooth muscle of each organ is distinctive from that of most other organs in several ways: dimensions, organization, response to stimuli, and functions.

In contrast to the skeletal muscle fibres, which are 20 times as large in diameter and thousands of times as long, the smooth muscle can maintain prolonged tonic contractions for hours with little use of energy [1]. In fact, the energy required to sustain smooth muscle contractions is only $\frac{1}{10}$ to $\frac{1}{300}$ of what is required to sustain the same tension of contraction in skeletal muscle. Moreover, the maximum force of contraction per unit section is often even greater than that of the skeletal muscle and the useful distance of contraction is more than $\frac{2}{3}$ of its stretched length. This allows smooth muscle to perform especially important functions in the hollow viscera, allowing for example the gut to change its lumen diameter from very large (10–15 cm) down to almost zero.

In contrast to the skeletal muscle activated exclusively by the nervous system, the contraction of smooth muscle can be stimulated by multiple types of excitations: neural, hormonal, thermal, and mechanical. The principal difference is that the smooth muscle membrane contains many types of receptors that can initiate or inhibit the contractile process.

It is customary to classify smooth muscle into multi–unit and single–unit smooth muscle. The most important characteristic of the multi–unit smooth muscle is that each fibre can contract independently from the others, and their control is exerted mainly by nerve signals. Ciliary muscle, iris of the eye, and piloerector muscles of the hair are examples of multi–unit smooth muscle, while single–unit or unitary smooth muscle can be found in the gut, bile ducts, ureters,

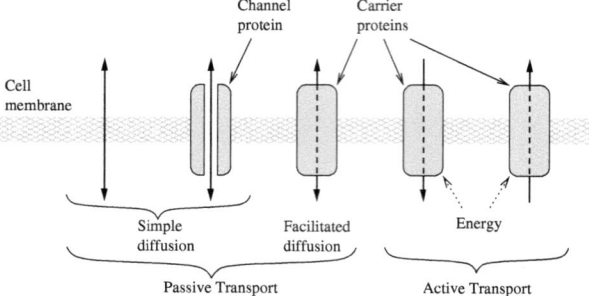

Figure 2.1: Transport pathways through the cell membrane.

uterus, and many blood vessels. This type of smooth muscle consists of a whole mass of hundreds to millions of muscle fibres that contracts together as a single unit.

In this chapter we focus our attention on the basic structure of cellular membrane and the main mechanisms permitting exchanges of chemicals across it. The electric potential of the membrane is the subject of a section, in which the principles of resting membrane potential and action potentials are described. Electric activity of the smooth muscle cell is treated as a particular case in the last part of this chapter.

2.1 Cell Membrane

Cell is the structural and functional unit of living organisms. It is capable by itself or by interacting with other cells of performing the fundamental functions of life. The organisms consisting of a single cell, such as bacteria, are called unicellular. An organism composed of many cells is called multicellular. The human body for instance is composed of an estimated 10^{14} cells.

Each cell is a self–contained and self–maintaining entity that can take in nutrients, convert them into energy, carry out specialized functions, and reproduce if necessary. Within the cell take place all vital functions of an organism and are stored the hereditary information necessary for regulating cell activities and generating new cells. The interior of a cell is separated from its surroundings by a protective coat, called cell membrane. The structure of the cell membrane consists almost entirely of a lipid bilayer with large number of proteins floating in the lipid, many of which penetrating all the way through. The lipid bilayer is hydrophobic, thus not miscible with the extracellular and intracellular fluid, which lies respectively outside and inside the cells. Even though a few substances can penetrate this bilayer and enter the cell or leave it, passing directly through the lipid substances itself, the cell membrane constitutes a barrier for the movement of many molecules between the extracellular and intracellular fluid compartments. Depending on the molecule, transport through the cell membrane (see figure 2.1), either directly through the lipid bilayer or through the transport proteins, occurs by two basic processes: the passive transport that does not consume chemical energy (ATP), and active transport that does consume chemical energy. The molecular structure of transport proteins interrupts the continuity of the lipid bilayer and therefore constitutes an alternate pathway through the cell membrane.

2.1.1 Passive Transport

The extracellular fluid contains a large amount of sodium and chloride but only a small amount of potassium, the exactly opposite of the intracellular fluid. Concentration asymmetries of ions between the two regions are extremely important to the life of the cell, and are often termed as concentration gradients. Passive transport allows free movement of chemicals across cell membrane without involving chemical energy, through a diffusion phenomenon [2].

Diffusion is the net movement of substances from an area of high concentration to lower concentration, following the concentration gradient. Substances that diffuse in one direction can also diffuse in the opposite direction but what is important is the net rate of diffusion of a substance in the desired direction. If and when the concentration gradient has been eliminated, no net exchange of material occurs. The most important factors that determines how rapidly a substance will move through the lipid bilayer are the membrane permeability, the concentration difference, the membrane electric potential, and the pressure difference.

Diffusion through the cell membrane is divided into two subtypes called simple diffusion and facilitated diffusion. Simple diffusion is the movement of molecules across the cell membrane, either passing directly through the lipid bilayer, especially if the diffusing substance is lipid–soluble, or via special channel proteins that are embedded within the cellular membrane, if the molecules are small enough. The protein channels are distinguished by two important characteristics: they are often highly selectively permeable to certain substances and many of them can be opened or closed by gates. Selective permeability for the transport of one or more specific ions or molecules results from the characteristics of the channel itself, such as its diameter, its shape, and the nature of the electric charges along its inner surface. On the other hand, gating of protein channels provides a means for controlling the permeability of the channels by opening and closing the pathway. The molecular conformation of the gate responds to the electric potential difference between the cell membrane in the case of voltage gating type, and to the presence or absence of a particular molecule in the case of chemical gating. The gate of a channel is either totally open or completely closed, the transition between which occur within a few millionths of a second. Under particular conditions, the channel may remain closed all or almost all the time, whereas under other conditions it may remain open all or most of the time. With in–between states the gates tend to snap open and closed intermittently, giving an average current flow somewhere between the minimum and the maximum.

Facilitated diffusion is the movement of molecules across the cell membrane via special carrier proteins that are embedded within the cellular membrane. The carrier protein aids passage of large or lipid–insoluble molecules through the membrane by binding chemically with them and shuttling them through the membrane in this form. Facilitated diffusion is a passive process that however transports molecules down the concentration gradient. Among the most important substances that cross the cell membrane by facilitated diffusion are glucose and amino acids.

2.1.2 Active Transport

A large concentration of a substance is often required in the intracellular fluid even though the extracellular fluid contains only a minute concentration, which is the case, for instance, of potassium ions. Conversely, sometimes it is important to keep the concentrations of other ions very low inside the cell even though their concentration in the extracellular fluid are great, e.g. sodium ions. Neither of these two effects could occur by simple diffusion because simple diffusion equilibrates the concentrations on the two sides of the membrane. Instead, some energy source is needed for the net movement against the concentration gradient. When molecules or ions are moved through the cell membrane from an area of low concentration to an area of higher concentration, thus against a concentration gradient, the process is called active transport [2].

Active transport is divided into two types according to the source of the energy used to cause the transport. They are called primary active transport and secondary active transport. In a primary active transport, the energy is directly coupled to the movement of the desired substance across the membrane, and derived directly from the breakdown of adenosine triphosphate (ATP) or some other high–energy phosphate compounds.

A primary active transporter universal to all cellular life is the sodium–potassium pump (Na/K–ATPase). It is a transport process that pumps sodium ions outward through the cell membrane and at the same time pumps potassium ions from the outside to the inside. When three sodium ions bind on the inside of the carrier protein and two potassium ions on the outside, the ATPase function of the protein becomes activated and cleaves one molecule of ATP, liberating a high amount of energy in its decomposition. This energy causes a conformational change in the protein carrier molecule, extruding the sodium ions to the outside and the potassium ions to the inside.

Secondary active transport uses the flow of one solute species from high to low concentration to move another molecule against its preferred direction of flow. The two main forms of secondary active trasport are co–transport and counter–transport. In co–transport both solutes move in the same direction across the membrane, while in counter–transport the solutes move in opposite direction. Primary active transport usually develops a large concentration gradient of sodium ions that always attempt to diffuse from the outside to the interior of cells. This concentration gradient represents a storehouse of energy, and under appropriate conditions it is also used to drive other substances through the cell membrane. The coupling mechanism required for sodium to pull another substance along with it, is achieved by means of a carrier protein in the cell membrane that serves as an attachment point for both the sodium ion and the substance to be transported. In sodium co–transport, the carrier protein presents both bindings on the external surface of the cell membrane, while in counter–transport the sodium binding is on the external surface and the other chemical binding is on the inside of the cell. A conformational change occurs in the carrier protein once they are both bound, and the energy gradient of the sodium ion causes both the sodium ion and the other substance to be transported together across the cell membrane, in the same direction in the case of co–transport and in opposite directions in counter–transport. Glucose is transported into most cells against large concentration gradients by the mechanism of sodium co–transport. In a similar way as for glucose, amino acids are transported into the cells, but using a different set of transport proteins. Five amino acids transport proteins have been identified, each of which is responsible for transporting one subset of amino acids with specific molecular characteristics.

2.1.3 Measuring Ionic Channel Current

A wealth of new information about ion channels resulted from the invention of the patch–clamp method in the 1970s. This technique is based on a simple idea and makes it possible to record ion current flow through single channel. A very small pipette, having a diameter of only 1 or $2\mu m$, is abutted against the outside of a cell membrane. Then suction is applied inside the pipette to pull the tiny area, or patch, of cellular membrane slightly into the tip of the pipette. This creates a seal where the edge of the pipette touches the cell membrane and no ions can flow between the pipette and the membrane. The result is a minute patch at the tip of the pipette through which all the ions that flow when a single ion channel opens can be observed. The resulting electrical current, though small, can be measured with an ultrasensitive electronic amplifier connected to the pipette.

Moreover, the small cell membrane patch at the end of the pipette can be torn away from the cell. The pipette with its sealed patch is then inserted into a free solution. This allows the ion concentration both inside the micropipette and in the outer solution to be altered as desired. Also, the voltage between the two sides of the membrane can be set as will (clamped to a given voltage). By varying the concentrations of different ions and the voltage across the membrane, one can determine the transport characteristics of the channel as well as its gating properties.

2.2 Membrane Potential

All cells are separated from their environment by their cell membrane, and electric potentials exist across the membrane of essentially all cells. In fact, because composed principally of a lipid bilayer, the cell membrane is a good insulator between the intracellular and extracellular electric environments. The potential difference between the inside and the outside of the cell is called the membrane potential [3].

Calcium, potassium, sodium, and chloride ions are the main ions most involved in the development of membrane potentials in smooth–muscle fibres as well as in nerve fibres. The concentration gradient of each of these ions across the membrane determines the voltage of the membrane potential and the degree of importance of each of the ions in determining the voltage is proportional to the membrane permeability for that particular ion.

2.2.1 The Nernst Equation

Ions tend to flow across the cell membrane from the side on which their concentration is higher to the side on which their concentration is lower. Because ions are electrically charged, their flow also depends on the potential across the membrane, i.e. the electric gradient. The net force acting on ions is determined both by the electrical and chemical gradients and is referred to as the electrochemical gradient or driving force. When electrical and chemical gradients are equal and opposite, no net diffusion of ions in either direction through the membrane is possible, and the potential level across the membrane is called the equilibrium or Nernst potential. The magnitude of this potential is determined by the ratio of the ion concentrations on the two sides of the membrane. The greater this ratio, the greater the tendency for the ions to diffuse in one direction, and therefore the greater is the Nernst potential. The following equation is called the Nernst equation and is used to calculate the Nernst potential for any univalent ion k at normal body temperature of 37°C:

$$E_k = \frac{RT}{zF} \ln \left(\frac{[k]_e}{[k]_i} \right)$$

where R is the universal gas constant, T the absolute temperature, z the valence of the ion k, F the Faraday constant, and $[k]_e$ and $[k]_i$ are the extracellular and intracellular ion concentrations respectively.

The difference between the transmembrane potential $V_m = (V_i - V_e)$ and the Nernst potential E_k of ion k gives the value of the driving force. Moreover, the sign of the driving force $(V_m - E_k)$ determines the direction of movement of the ionic current across the membrane.

When the three ions, potassium, sodium, and chloride are involve in the process, the Goldman–Hodgkin–Katz equation has to be used to determine the equilibrium potential.

2.2.2 Resting Membrane Potential and Action Potentials

Cells in general maintain a small potential across their membrane. In the normal or resting state the inside of the cell is negative with respect to the outside (typically about -50 to -60 millivolts). The potential arises from the differences in concentration across the cell membrane of the electrolyte ions, namely calcium, potassium, sodium, and chloride. The uneven distribution of electrically charged ions is established in part by the properties of the membrane, and in part by the primary active transporters. When there is no net movement of ions across the membrane, the equilibrium is reached and the so called resting membrane potential is established.

Some cells, such as nerve cells and muscle cells, are excitable. They can be electrically stimulated and are capable of transmitting this electric excitation called the action potential. It is an all–or–none electric pulse that propagates rapidly over the cellular membrane. It is generated by a local depolarization across the membrane followed by a repolarization. The alteration in the calcium and potassium permeabilities is primarily responsible for the action potentials in the smooth muscle cells (sodium and potassium in nerve fibres). The necessary actors in causing both the depolarization and repolarization of the membrane during an action potential are the voltage–gated calcium and potassium channels. The calcium channels are also called slow channels because they activate more slowly than sodium channels. However they remain open much longer. This accounts in a large measure for the slow action potentials of the smooth muscle fibres. Another important feature of the calcium entry into the cell during the action potential is that the same calcium acts directly on the contractile mechanism of the smooth muscle.

An action potential elicited at any point on an excitable membrane usually excites adjacent portions of the membrane, resulting in propagation of the action potential. An excitable membrane has no single direction of propagation, therefore the action potential will travel in both directions away from the stimulus until the entire membrane has become depolarized.

Moreover, cell membranes of two neighboring fibres are joined by many gap junctions through which ions can flow freely from one cell to the next and cause the muscle fibres to contract together. Gap junctions act as low resistance pathways for the rapid spread of electrical signals throughout the tissue.

2.2.3 Measuring Membrane Potential

Resting membrane potential is measured using an appropriate voltmeter that is capable of measuring very small voltages between two microelectrodes. The first electrode is implanted through the cell membrane to the interior of the cell, and the other one is placed in the extracellular

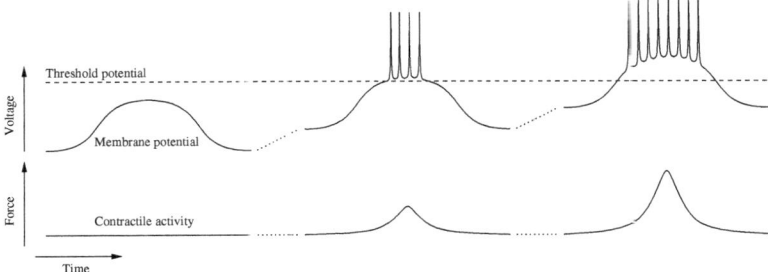

Figure 2.2: Membrane potentials of the intestinal smooth-muscle.

fluid. The same assessment can also be used to measure monophasic action potential. When the action potentials spreads down the cell membrane, the changes in the potential inside the cell are observed.

Biphasic recordings, the most usual method to assess action potentials, are obtained by placing two electrodes on the outside of the cell at a specific distance and recording the potential variations on the surface.

The monophasic and biphasic methods of measuring the membrane potentials are simple in theory but often difficult in practice because of the small size of the cells.

2.3 Slow Waves and Spike Potentials

Gastrointestinal motility is governed by myogenic, neural, and chemical control mechanisms. The myogenic control system manifests itself by periodic depolarizations of the smooth-muscle cells that constitute autonomous electrical oscillations called the electrical control activity or slow waves. These partial depolarizations of the membrane potential are fluctuations of 5 to 15 mV that spread to adjacent sections of the muscle on long distances. The frequency of this intrinsic phenomenon depends on the section of the digestive tube: about 3 times per minute in the stomach, as much as 12 in the duodenum, 8 or 9 in the terminal ileum, and about 1 to 5 in the colon. Although slow waves are not action potentials and usually do not cause muscle contraction themselves (except in the stomach), most gastro-intestinal contractions occur rhythmically, and this rhythm is determined mainly by the frequency of the slow waves. The role of the slow waves is to regulate the contractile excitability of the smooth-muscle cells, since the cells may contract only when depolarization of the membrane voltage exceeds an excitation threshold. The normal spontaneous amplitude of the slow wave depolarization does not exceed this excitation threshold except when neural or chemical excitation is present. The myogenic system affects the frequency, the direction, and the velocity of the contractions. It also affects the coordination or lack of coordination between adjacent segments of the gut wall.

Slow waves control the appearance of another type of electric waves named electrical response activity or spike potentials, which are the real cause of the contractions. Spike potentials are true action potentials characterized by intermittent bursts of rapid electrical oscillations. They occur during the depolarization plateau of the slow waves if a cholinergic stimulus is present, and they are associated with muscular contractions [4]. Thus, the neural and chemical control systems determine whether contractions occur or not, but when contractions take place, the myogenic control system determines the spatial and temporal patterns of the contractions.

In addition to the slow waves and spike potentials, the voltage level of the resting membrane potential can change. Under normal conditions, the average of the resting membrane potential is about -55 mV, but multiple factors can change this level. When the potential becomes more positive, which is called the depolarization of the membrane, the muscle fibre becomes more excitable. In the opposite case, when the potential becomes more negative, which is called hyperpolarization, the fibre becomes less excitable.

2.4 Conclusion

The brief overview of the major elements of the gastro–intestinal smooth–muscle cell exposed in this chapter has to be seen as a glossary. In fact, the terms, the principles, and the processes introduced over this chapter will be widely used in the following of the dissertation or will be helpful for a better comprehension.

In this chapter, a description of the principal differences between muscle fibres of the body has been given, as well as a presentation of the principal characteristics and properties of the gastro–intestinal smooth–muscle tissue.

The structure of the cellular membrane and its specificity have first been discussed, in order to introduce the different processes involved in the substance transport across the membrane. The basic physical and physiological principles of the passive and active transport have been exposed. The patch–clamp, a powerful method for measuring the ionic channel currents, has been presented. This technique is useful for the mathematical description of the behaviour of ionic channels in cellular models, as we will see in the second part of the dissertation.

The description of the nature of the membrane potential has been given, and its components, the resting membrane potential and the action potentials, have been exposed. The equilibrium potential for a specific ion generated by the electrochemical gradients through the membrane is estimated using the Nernst equation, also introduced in this chapter. The monophasic and biphasic techniques for measuring the membrane potentials and their propagations along the membrane have been presented.

Specific aspects of the electrical activity of the gastro–intestinal smooth–muscle cell have been treated, and the slow waves and spike potentials have been described.

Chapter 3

The Human Digestive System

The task of the gastro–intestinal system (GIS) is to provide the body with water and nutriments, in order to allow the maintenance of the organism normal functions. The constitutive specialized organs work in association so as to achieve this goal. At its simplest, the GIS can be considered as a portal for food from the environment to the circulatory system. The ingested food is reduced to very simple molecules by a combination of mechanical and enzymatic degradation, transported across the epithelium lining the intestine into blood, and subsequently used for the metabolic purposes of all cells in the organism. In order to accomplish such a transfer, the ingested food is moved along the alimentary canal at an appropriate rate for digestion, absorption, storage, and excretion. To fulfil the various requirements of the system, each organ has developed one or more specific functions: some limited to simple passage of food (oesophagus), others to storage and mixing of food (stomach) and others to digestion and absorption (small intestine).

In the first section of this chapter, we present an anatomical and structural overview of the human digestive system, while the basic functional principles in the entire alimentary tract are discussed in the following section. The last section is dedicated to the control of the GIS.

3.1 Structure

From the anatomical point of view [5], the digestive system may be approximated as a tube running from the mouth to the anus with associated accessory digestive organs, such as the salivary glands, the liver, and the pancreas. Its average length in humans is approximately 6 to 8 m, depending on the size and age of the person. Local occlusions, called sphincters, segment the tube into several functional segments (figure 4.1), each one responsible for a particular task in the complex digestive process

Ingested food begins its decomposition in the mouth, where it is broken down mechanically by the chewing movements, and saliva is added as a lubricant. The oesophagus acts as a conduit for food from the pharynx to the stomach. Its role in digestion is simple, but routine exposition to rough and abrasive foodstuffs force it to possess a stratified epithelium resistant to trauma. The two extremities of the oesophagus are bounded by sphincters known as the upper and lower oesophageal sphincters. Both are constantly closed except during swallowing, which prevents constant entry of air from the oral cavity or reflux of stomach contents. The stomach is an expanded section of the digestive tube, where food is mixed with gastric secretions. It has three muscle layers: the outermost is the longitudinal, the middle is the circular, and the innermost is

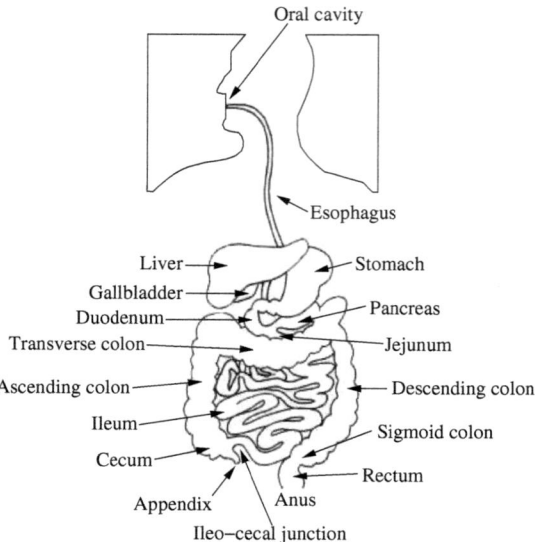

Figure 3.1: Segments of the GIS: oral cavity, oesophagus, stomach, small intestine (subdivided into duodenum, jejunum and ileum), colon (subdivided into caecum and ascending, transverse, descending and sigmoid colon), sigmoid, rectum, and anus.

the oblique. The most distal and narrow section of the stomach is termed the pylorus, through which passes the liquefied food, called chyme, into the small intestine. The small intestine is the longest section of the digestive tube (5 to 7 m–long) and consists of three segments forming a passage from the pylorus to the large intestine: the duodenum, the jejunum, and the ileum. The duodenum is a short C–shaped section, about 18 cm long, that receives secretions from the pancreas and liver via the pancreatic and bile ducts. The jejunum constitutes one–third of the small gut in man, and the ileum the remaining two–thirds. The major differences between the jejunum and the ileum are functional in nature, and are related mainly to their absorption characteristics. The chyme is propelled by a process of muscular contractions called peristalsis along the small intestine and through the ileocaecal valve to the large intestine. The ileocaecal valve prevents material from flowing back to the small intestine and is positioned approximately 5 cm from the proximal end of the large intestine. The caecum is the first part of the large intestine, measuring roughly 1 m long, which consists also of the colon and the rectum. The colon is sacculated and the longitudinal smooth muscle is concentrated in three bands, the taeniae. It consists of the ascending colon on the right side, the transverse colon between the hepatic and the splenic angle, the descending colon on the left side, and the sigmoid (meaning S–shaped). The rectum is the final portion of the large intestine, terminating in the anus.

From the mouth to the anus, the wall of the digestive tube is composed of four basic layers or tunics (figure 3.2), from the innermost outwards are the mucosa, the muscular layer (muscularis)

Section 3.1 Structure

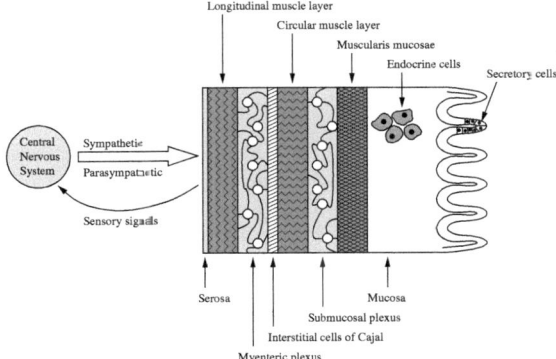

Figure 3.2: The wall of the tube consists of three principal layers which, from the innermost outwards are the mucosa, the muscular layer (muscularis), and the serosa (or adventitia).

and the serosae (or adventitia), which all serve specific functions.

Among the four tunics, the mucosa is the most variable in structure and function, endowing the tube with an ability to perform diverse and specialized digestive tasks along its length. Of critical importance in this regard are the epithelial cells, namely enterocytes, that cover the mucosa and are thus in direct contact with the lumen. The single–layer of enterocytes represent the interface between the outside and the inside of the body. The mucosal surface is greatly increased by the elongation and multiple bendings of the tube, the formation of multiple foldings (plication and villosities) and the sub–microscopical folding of the enterocyte membrane facing the lumen.

The tunica submucosa, immediately beneath the mucosa, is a layer of loose to dense connective tissue containing blood and lymphatic vessels. The submucosa also contains the submucous plexus, a critical component of the digestive tract's nervous system which provides nervous control to the mucosa.

The tunica muscularis endows the digestive tube with an ability to be motile. In the digestive tube, this tunic consists of two to three thick layers of smooth muscle. Muscle fibres in the outer layer have a longitudinal orientation, whereas those in the inner layer are aligned circularly (circumferential). In some segments of the gastro–intestinal tract, an innermost layer where fibres are obliquely disposed is present. The combination of circular and longitudinal smooth muscle gives the tube an ability to perform complex movements that squeeze and propel ingesta in the lumen. Between the inner circular and outer longitudinal layers of smooth muscle is another critical component of the digestive tract's nervous system, the myenteric plexus.

The tunica serosa consists of a thin layer of loose connective tissue and covers the tube within the abdominal cavity, where it allows for an unrestrained sliding of moving intestinal loops. The serosa on each side of the tube fuses together to form a suspensory structure called mesentery, which houses vascular and nervous supplies to the digestive tract and is continuous with the lining of the cavity, the peritoneum. Outside the peritoneal cavity, the serosa fixes the tube to the neighboring tissues.

At first glance, the structure of the small intestine is similar to other regions of the digestive tube, but when examined closely, the lumenal surface of the small intestine appears similar to velvet because it is covered by millions of small projections called villi which extend about 1 mm into the lumen. The other fundamental structures constituting the small intestinal mucosa are the crypts, the moat–like invaginations of the epithelium around the villi. Villi are only the most obvious feature of the mucosa which houses a dynamic, self–renewing population of epithelial cells, which take up nutrients from the lumen and transport them into blood, fulfilling the basic function of the digestive system. In addition, the apical (exposed) surface of the epithelial cells of each villus is covered with microvilli. Thanks largely to these, the total surface area of the intestine is almost 200 square meters, about the size of the singles area of a tennis court and some 100 times the surface area of the exterior of the body. If the small intestine is viewed as a simple pipe, its lumenal surface area would not be greater than one half of a square meter.

3.2 Functions

The mission of the GIS is to provide energy from the ingested food for building and sustaining cells. When we eat food such as bread, meat, or vegetables, they are not in a form that the body can use as nourishment. Thus, the GIS proceeds like an assembly line, or more properly, a disassembly line. Food and drink macromolecules (proteins, fats, and starch), which cannot be absorbed in their original state, are decomposed into smaller molecules of nutrients (amino acids, fatty acids and glucose) before they are absorbed through the wall of the tube, into the circulatory system to be carried to cells over the body.

To achieve its goal, the digestive system requires movement of food through the alimentary tract, secretion of digestive juices, digestion of the food, absorption of the digestive products, water, and various electrolytes; defense barrier against dangerous agents. These five fundamental functions that take place in the digestive system are presented in this section.

3.2.1 Motility

Motility is the function of the digestive system consisting in the ability of moving spontaneously, independently, and coordinately the muscles composing it. Except for the muscles of the mouth and the first section of the oesophagus, all the muscles in the wall of the digestive tube are smooth muscles. Indeed, the patterns of motility seen in the gut have the characteristics of smooth muscle activities.

Muscular movements in the gastro–intestinal tract have the purpose of assisting, favoring, and facilitating digestion and absorption. Each segment of the tube is specialized in one or more tasks so that the ingested aliments are mechanically fragmentated, mixed with digestive secretions, maintained in a permanent contact with absorptive cells, propulsed in the oral–aboral direction, stored, and finally expulsed.

Two fundamental types of movements occur in the gastro–intestinal tract [5]–[6]: propulsive movements and mixing movements. Propulsive movements cause food to move along the tract at an appropriate rate for digestion and absorption. The basic propulsive movement of the gastro–intestinal tract is peristalsis, which consists of a contractile ring appearing around the tube and then moving along like a wave on a surface of water. The propagation of the peristaltic wave is characterized by a pressure increase in the stimulated zone, and at the same time a muscle relaxation at the successive portion of tube, in order to generate a pressure gradient. Peristalsis, theoretically, can propagate in either direction, but it normally dies out rapidly in the oral direction while continuing for a considerable distance analward. If the ingested materials were simply propelled through the digestive tube, digestion and absorption would be very poor. The mixing movements help the combination of digestive enzymes with the ingesta and facilitate the absorption of nutrients by the epithelial cells. Alternating contraction and relaxation of the smooth muscle in the tube provides effective mixing of its contents.

The first major muscle movement occurs when food or liquid is swallowed. Swallowing is initiated voluntarily in the mouth and is pursued systematically in the pharynx. The involuntary mechanism is specifically organized to promote the passage of food from the pharynx through the oesophagus to the stomach. The oesophagus normally exhibits two types of peristaltic movements: primary peristalsis and secondary peristalsis. Primary peristalsis is the strong peristaltic wave that begins in the pharynx and spreads along the oesophagus in about 8 to 10 s. If the primary peristaltic wave fails to deliver all the food that has entered the oesophagus into the stomach, a secondary peristaltic wave is initiated at the point of distension and they continue until all the food is emptied into the stomach. At the junction of the oesophagus and stomach, there is a ring–like valve closing the passage between the two organs. However, as the food approaches the closed ring, the surrounding muscles relax and allow the food to pass.

Large quantities of food and liquid can be stored in the stomach until they are processed in the small intestine. The first motor function requires the muscle of the stomach to relax and accept large volumes of swallowed material. The second mechanical task is to mix the ingesta with gastric juice produced by the stomach until it forms a semifluid mixture called chyme. Finally, it has to empty its contents slowly into the duodenum, at a rate suitable for proper digestion and absorption by the small intestine.

Gastric contraction waves are initiated by the basic electrical rhythm and consist of electrical slow waves that occur spontaneously in the stomach wall, at a rate of about three per minute.

The movements of the small intestine can be divided into mixing contractions and propulsive contractions, although the separation is artificial because essentially all movements of the small intestine cause at least some degree of both mixing and propulsion.

Mixing movements in the small intestine are characterized by localized concentric contractions spaced at intervals along the tube. Each set of contractions causes segmentation of the small intestine, and as one set of ring–like contractions relaxes, a new one begins. The maximum frequency of the segmentation contractions is determined by the frequency of the basic electrical rhythm in the intestinal wall, namely slow waves. In the duodenum and proximal jejunum, the maximum frequency of the segmentation contractions is, under extreme conditions of stimulation, about 15 to 20 per minute, while in the terminal ileum it is usually 8 to 10 contractions per minute.

Moreover, chyme is propelled through the small intestine by peristaltic waves that can occur in any part of the small intestine. Propulsion waves move principally analward at a velocity of 0.5 to 2.0 cm per second, much faster in the proximal intestine and much slower in the terminal intestine. They normally are very weak and usually die out after traveling only 3 to 5 cm, very rarely farther than 10 cm. The movement of the chyme is then very poor, so poor in fact that 3 to 5 hours are required for chyme to pass from the pylorus to the ileocaecal valve. This means a net movement of the chyme along the small intestine averaging only 1 cm per minute.

In the large intestine take place the terminal phase of digestion. Its functions are absorption of remaining water and electrolytes from ingesta, principally in the proximal half, and storage of fecal matter until it is expelled in the distal half. Intense movements are not required for these functions, thus the large intestine motility is normally slowed down. Yet, in a sluggish manner, the movements still have characteristics similar to those of the small intestine and can be divided once again into mixing movements and propulsive movements.

Segmentation movements occur in the large intestine by constriction of the circular muscle in order for the lumen to become almost occluded and, at the same time, contraction of the longitudinal muscle, which is aggregated into three longitudinal strips called the teniae coli. These combined contractions of the circular and longitudinal strips of muscle force the unstimulated portion of the large intestine to bulge outward as a bag–like sac called haustration. The haustral contractions, once initiated, usually reach peak intensity in about 30 s and then disappear during the next 60 s.

Peristaltic waves of the type seen in the small intestine only seldom occur in most parts of the colon. Instead, most propulsions occur by the slow analward movement of the haustral contractions and mass movements. Mass movements or giant migrating contractions are a distinct pattern of motility not seen elsewhere in the digestive tube. They are like very intense and prolonged peristaltic contraction that strips a segment of the large intestine (20 cm or more), with the role of sweeping residual undigested material in the distal direction through the digestive tube. The contraction force develops progressively for about 30 s, and they relaxes during the next 2 to 3 min before the next mass movement. The whole series of the mass movement usually persists for only 10 to 30 min, and usually occur only 1 to 3 times each day, most abundantly for about 15 min during the first hour after breakfast.

Much of the propulsion in the caecum and ascending colon results from the slow but persistent haustral contractions, while from the beginning of the transverse colon to the sigmoid, mass movements mainly take over the propulsive role. When a mass movement forces feces into the rectum, a reflex contraction of the rectum and relaxation of the anal sphincters are normally initiated.

3.2.2 Secretion

Secretion [7] is the function of digestive glands which consists, in addition to the luminal contents, of watery solutions containing ions, biliary components, and digestive enzymes (figure 3.3). Mixed with the ingesta, these substances activate a cascade of chemical reactions through which complex swallowed nutrients are decomposed into simpler molecules. The other primary function of secretory glands, more precisely mucous glands (and cells), is to provide mucus for lubrication and protection of all parts of the alimentary tract, from the mouth to the anus.

Most digestive secretions are formed only in response to the presence of food in the alimentary tract, and the quantity secreted in each segment of the tract is almost exactly the amount needed for proper digestion. Furthermore, in some portions of the gastro–intestinal tract, even the type of enzymes and other constituents of the secretions change depending on the type of food present.

The salivary glands, situated in the mouth, are accessory digestive glands that produce between 0.5 and 1.5 l a day of a fluid secretion called saliva. There are three major pairs of salivary glands: the parotid gland, the submandibular gland, and the sublingual gland. Saliva dissolves food particles so that they be tasted, plays a particularly important role in maintaining

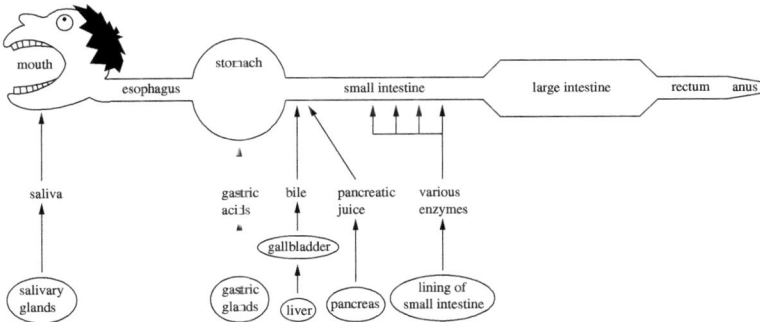

Figure 3.3: Schematic view of the digestive tract. The digestive glands and their secretions are also showed in their natural order of interaction in the digestive process.

the health of oral tissues by washing away and destroying pathogenic bacteria, contains enzymes that acts with foodstuffs, and joins the mucus lubricating the pharynx to facilitate swallowing. Mucus is also secreted by oesophagus lining principally to provide lubrication for swallowing.

The next set of digestive glands is located in the stomach lining. Millions of gastric glands produce together between 400 and 800 ml of gastric juice at each meal, permitting the dissolution of the ingested food. The secretion of mucus by the lining cells of the stomach make its mucosa fortunately resistant to the extreme acidity of the gastric juice, which is triggered by the sight, smell or taste of food, so that the stomach is prepared when the food arrives.

After the stomach empties the food and juice amalgam into the small intestine, the juices of two other digestive organs (pancreas and liver) are mixed with them to continue the process of digestion. The pancreas is a retroperitoneal organ that produces pancreatic juice containing bicarbonate ions, water, and enzymes for digesting carbohydrates, proteins, and fats. Its secretions are drained via the main pancreatic duct directly into the duodenum, and most abundantly in response to the presence of chyme in the upper portions of the small intestine.

The liver is the centre of metabolic activity in the body and its major role in the digestive process is to produce between 0.6 and 1.2 l a day of the digestive juice named bile. The bile does not contain enzymes, but plays an important role in fat digestion and absorption. The bile secreted continually by the liver cells is normally stored between meals in the gallbladder, a small sac situated just below the liver and attached to it by tissues. At mealtime, when food leaves the stomach, a secretion causes the gallbladder to contract and release its contents into the duodenum via the bile ducts

In crypts located between villi on the entire lining of the small intestine, there are cells secreting mucus that lubricate and protects the intestinal surface, and enterocytes secreting about 1.8 l of water and electrolytes. Over the surface of villi, a large number of enterocytes reabsorb electrolytes and water along with the end products of digestion. The circulation of fluid from the crypts to the villi supplies a watery vehicle for absorption of substances from the chyme as it comes in contact with the villi. The mucosa of the large intestine has many crypts but no villi, and the great preponderance of secretion is mucus.

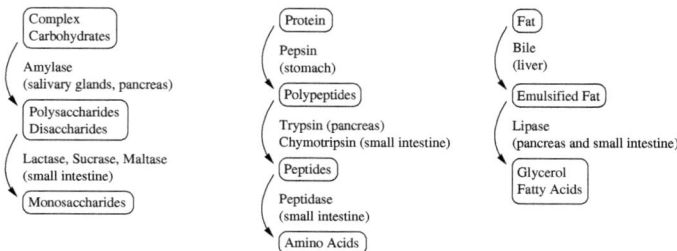

Figure 3.4: Digestion steps of the carbohydrates, fats, and proteins in the human body. Enzymes and chemicals interacting in the process of decomposition so as the secretion glands of the digestive system are indicated in the three diagrams.

3.2.3 Digestion

The foods on which the body lives, with the exception of small quantities of substances such as vitamins and minerals, can be classified as carbohydrates, fats, and proteins. They cannot be absorbed in their natural forms through the gastro–intestinal mucosa. They have to be digested before. Digestion [8] is the process of making food absorbable to the body by breaking it down and dissolving it into simpler chemical compounds.

Mechanical digestion (chewing, churning or grinding) is a physical process that breaks food into smaller pieces, without changing them chemically. The importance of mechanical digestion is that by breaking food into smaller pieces it increases the surface area on which digestive enzymes can react during chemical digestion.

Chemical digestion is a process that decomposes nutrients into simpler molecules through the action of enzymes and chemicals secreted by the digestive system (figure 3.4).

The first secretion involved in this process is produced by salivary glands and is added to food at its entry into the gastro–intestinal tube. The saliva contains an α–amylase enzyme that starts the digestion of starch. Starch is a large polysaccharide and is with sucrose and lactose one of the three major sources of carbohydrates in the normal human diet. The specific enzyme separates starch into maltose and other small polymers of glucose.

The epithelium of the stomach forms deep pits, and three different types of cells are at different locations below them. The cells at the base of these pits are chief cells, responsible for production of pepsinogen, an inactive precursor for the enzyme pepsin, which degrades proteins formed from amino acids bounded together by peptide bond. The secretion of the enzyme in its inactive form prevents self–digestion of the stomach cells. Further up the pits, parietal cells produce hydrochloric acid, which is a fairly strong acid that helps to liquefy foodstuffs and especially activates pepsinogen into pepsin. The enzyme starts the digestion of proteins by breaking long polypeptides chains into shorter ones. A protein called intrinsic factor is that is essential for the absorption of vitamin B12 by the intestine is also produced by parietal cells of the stomach. Near the top of the pits, closest to the contents of the stomach, there are mucus producing cells that help protect the stomach from self–digestion.

Bile is a liquid secreted by the liver and stored in the gallbladder that especially assists in the digestion of fats. The bile acids dissolve the fat into the watery contents of the intestine through a process named emulsification, so that the water–soluble digestive enzymes can act on small size globules surfaces of triglycerides.

The pancreas produces a juice that contains a wide array of enzymes to break down the carbohydrate, fat, and protein in food. Pancreatic secretions, like saliva, contain a large quantity of α–amilase almost identical in its function to the α–amylase of saliva, but several times as powerful. Starches are totally converted into maltose and other small polymers of glucose. Lipase is another enzyme contained in the pancreatic juice that is essential for the digestion of fats. Most protein digestion occurs principally under the influence of the enzymes of pancreatic secretions. The two major proteases the pancreas excretes are trypsinogen and chymotrypsinogen, which are inactivated forms of trypsin and chymotrypsin. Once released in the intestine, the enzyme enterokinase present in the intestinal mucosa activates a cascade of chemical reactions so as to activate the proteases. Trypsin and chymotrypsin are able to break some peptidic links in protein molecules, thus it is important that enzymes of the pancreatic juice not become activated until they have been secreted into the intestine because they would digest the pancreas itself. Pancreatic secretions contain also bicarbonate ions in order to buffer the acidic chyme that the stomach churns out. After the fat is dissolved, it is digested by the pancreatic lipase, the most important enzyme for the digestion of triglycerides, which are transformed into glycerol and fatty acids.

Much of the digestion takes place on the surface of small intestinal epithelial cells, where enzymes that complete the digestion of peptides and disaccharides are produced. In the cell membrane of each microvillus are located multiple peptidases, the enzyme responsible of the splitting of polypeptides into tripeptides and dipeptides or finally to amino acids. The transformation of tripeptides and dipeptides ends in the enterocyte cytoplasm, where they are digested by specific peptidase to obtain amino acids. The enterocytes lining the villi of the small intestine contains also the three other enzymes, maltase, lactase, and sucrase, which are capable of splitting the disaccharides, maltose, lactose, and sucrose respectively as well as the other small glucose polymers into their constituent monosaccharides.

$$\text{maltose} \xrightarrow{maltase} 2 \times \text{glucose}$$
$$\text{lactose} \xrightarrow{lactase} \text{glucose} + \text{galactose}$$
$$\text{sucrose} \xrightarrow{sucrase} \text{glucose} + \text{fructose}$$

Maltose and other small polymers split all into molecules of glucose, while lactose splits into a molecule of galactose and a molecule of glucose, and similarly sucrose splits into a molecule of fructose and a molecule of glucose.

3.2.4 Absorption

Gastro–intestinal secretions initiate a cascade of chemical reactions through which complex swallowed nutrients are decomposed into simpler molecules capable of crossing the mucosa barrier and entering the blood and lymphatic circulation. The term of absorption [8] denotes the transport of water, electrolytes (sodium, chloride, potassium, etc.) and essentially all dietary organic molecules (monosaccharides, amino acids, and fatty acids) from the lumen across the epithelium. It consists each day of several hundred grams of carbohydrates, 100 or more grams of amino acids, 50 to 100 g of ions, and 7 to 9 l of water. There are two ways for transport across the epithelium of the gut: the paracellular transport across tight junctions between enterocytes (epithelial cells), and the transcellular transport across the plasma membrane of the enterocytes. Some molecules, water for instance, are transported by both routes. In contrast, the tight junctions are impermeable to large organic molecules from the diet, which are transported exclusively by the transcellular route.

The small and large intestine intestine must absorb massive quantities of water. A normal person ingest roughly 1 to 2 l of dietary fluid every day. Moreover, another 6 to 7 l of fluid is received by the small intestine daily as secretions from salivary and stomach glands, pancreas, liver and the small intestine itself. This becomes to a total of about 8 l. Approximately 80% of this fluid is absorbed in the small intestine and the remaining in the large intestine. The absorption of water from the lumen always occurs by osmosis. In particular, an efficient electrochemical process establishes a sodium gradient between the lumen and the intercellular space making paracellular transport of water possible. Sodium is absorbed from the lumen into the cell by several mechanisms and rapidly exported by active transport from the cytoplasm via sodium–potassium pumps (embedded in the basolateral membrane) into the small intercellular spaces between adjacent enterocytes. Water diffuses from the lumen into the paracellular spaces in response to the osmotic gradient established by elevated concentration of sodium, and finally flows with sodium into capillary blood within the villus.

Monosaccharides are an important source of energy but only rarely found in normal diets, they are derived by enzymatic digestion of more complex carbohydrates within the digestive tube. Absorption of glucose, galactose and fructose entails transcellular transport from the intestinal lumen, across the epithelium and into blood. Glucose is taken into the enterocyte by a sodium co–transport mechanism. The sodium–dependent hexose transporter is a molecule that carries both sodium and glucose into the cell, and will not carry either of the two alone. The essence of transport by the sodium–dependent hexose transporter involves a series of conformational changes induced by the binding and releasing of sodium and glucose. The transporter is initially oriented towards the lumen and is capable of binding sodium, but not glucose. When sodium binds, it induces a conformational change of the transporter molecule, which opens the glucose–binding pocket. Afterwards glucose binds and the transporter reorients in the membrane such that the pockets holding sodium and glucose are moved inside the cell. Sodium is dissociated into the cytoplasm, causing glucose binding to destabilize and dissociate itself as well. The unloaded transporter reorients back to its original, outward–facing position, waiting for a new cycle. It is the low concentration of sodium inside the cell, generated by active transport of sodium through the basolateral membranes of the enterocyte, that drags the sodium to the interior of the cell along with the glucose. Glucose is transported out of the enterocyte through another transporter in the basolateral membrane and then diffuses into capillary blood within the villus. Galactose is transported by almost exactly the same mechanism as glucose. On the other hand, fructose is not concerned by the sodium co–transport mechanism. It enters the cell from the intestinal lumen via facilitated diffusion through another transporter.

Most proteins of the diet consist of giant molecules that must be digested by enzymes before they can be absorbed through the luminal membranes of the intestinal epithelial cells (the small intestinal enterocytes) in the form of dipeptides, tripeptides, and a few free amino acids. The mechanism by which amino acids are absorbed is conceptually identical to that of glucose. The energy for most of this transport is supplied by a sodium co–transport mechanism. Peptide or amino acid molecules bind with a specific transport protein in the enterocyte membrane only after binding sodium. The sodium ion then moves down its electrochemical gradient to the interior of the cell and pulls the amino acid or peptide along with it, followed by the reorientation of the transporter to the original form. This is called co–transport or secondary active transport of the amino acids or peptides. A few amino acids do not require this sodium co–transport mechanism but are instead transported by facilitated diffusion.

Fats are digested to form monoglycerides and free fatty acids. Both immediately enter the entherocyte by simple diffusion across the membrane. Once inside the enterocyte, they are re-combined to the endoplasmic reticulum to synthesize triglycerides. Beginning in the endoplasmic reticulum and continuing in the Golgi, triglyceride is packaged with cholesterol, lipoproteins and other lipids into particles called chylomicrons. Transport into the circulation is particularly different from that of sugars and amino acids. From the basolateral surfaces of the enterocytes, via the lymphatic vessel that penetrates into each villus, the chylomicrons diffuses rapidly into blood.

3.2.5 Immuno–Defense

The lumen of the gastro–intestinal tract is outside of the body and much of it is heavily populated with potentially pathogenic microorganisms. The gastro–intestinal tract contains an immensely complex ecology of microorganisms. A typical person can harbour more than 500 distinct species of bacteria, representing dozens of different lifestyles and capabilities. Bacterial populations in the large intestine can also be functional, digesting carbohydrates, proteins and lipids that escape digestion and absorption in the small intestine. The composition and distribution of this menagerie varies with age, state of health and diet. The number and types of bacteria in the gastro–intestinal tract vary also by region. In healthy individuals the stomach and proximal small intestine contain few microorganisms, largely as a result of the bacteriocidal activity of gastric acid. Moreover, together with the useful nutrients, undesired matters such as bacteria, viruses, and toxins are also ingested. Digestive mucosa is colonized by enormous amounts of immuno–competent elements that represent a real defense barrier against potentially dangerous agents. Immuno–defense is not a properly digestive function but is vital for the good performance of the digestive system.

3.3 Control

The control of the digestive functions is achieved by means of a complex combination of nervous, mechanical, and hormonal regulation signals. They are originated within the enteric nervous system and enteric endocrine system and can be modulated by the central nervous system.

3.3.1 The Enteric Nervous System

The gastro–intestinal tract has its own nervous system called the enteric nervous system [6]–[9]. It lies within the wall of the gut, covering the whole GIS. The scale and complexity of the enteric nervous system is immense. Beginning in the oesophagus and extending all the way to the anus, it contains as many neurons as the spinal cord, i.e. about 100 millions.

The principal components of the enteric nervous system are two networks of neurons: the myenteric plexus and the submucosal plexus. The myenteric plexus is the outer plexus lying between the longitudinal and the circular muscle layer in the tunica muscularis. It controls mainly the gastro–intestinal movements and exerts control over the motility function. The submucosal plexus is the inner plexus and, as its name implies, is buried in the submucosa. Its principal role is in sensing the environment within the lumen, regulating gastro–intestinal blood flow and controlling epithelial cell function.

Myenteric and submucosal plexuses are interconnected. Moreover, parasympathetic and sympathetic fibres connect the enteric and the central nervous systems. The enteric nervous system can and does function autonomously, but normal digestive functions require communication links between the enteric and the central nervous system. Through these cross connections, the gut can provide sensory information to the central nervous system, and the central nervous system can affect the gastro–intestinal functions. In general, sympathetic stimulation causes inhibition of gastro–intestinal motor activity and secretion, and a contraction of gastro–intestinal sphincters and blood vessels, while parasympathetic activity typically stimulates the digestive activities.

Within the enteric plexuses are three types of neurons: the sensory neurons, the motor neurons, and the interneurons. Sensory neurons receive information from sensory receptors in the mucosa and muscle. Different sensory receptors have been identified in the mucosa that respond to mechanical, thermal, osmotic and chemical stimuli.

Sensory receptors in the mucosa are able to "taste" the lumenal content, while sensory receptors in muscle are sensitive to mechanical stretch and tension. Collectively, enteric sensory neurons compile a comprehensive battery of information on the gut contents and the state of the gastro–intestinal wall.

Motor neurons control the gastro–intestinal motility and secretion, as well as the absorption acting directly on a large number of effector cells, including the smooth muscle, secretory exocrine and endocrine gastro–intestinal cells.

Interneurons are mainly responsible for integrating information from the sensory neurons and providing it to programming enteric motor neurons.

In addition, there is a non–neural mechanism controlling the gastro–intestinal motility, based on a self–generated, rhythmic fluctuation of the electric potential of the smooth muscle cells. In fact, slow waves are generated by special pacemaker cells, called interstitial Cajal cells. Interstitial cells are interconnected and connected to muscle cells by intercellular connections, named gap junctions.

3.3.2 The Enteric Endocrine System

The other system that controls the digestive functions is the endocrine system [6], with its secreting hormones. Let us recall that hormones are chemical messengers secreted into blood that modify the physiology of the target cells. A target cell for a particular hormone is a cell that has receptors for that hormone and reacts to it.

Digestive function is affected by hormones produced in many endocrine glands, but the leading control is exerted by hormones produced in the gastro–intestinal tract that is the largest endocrine organ in the body. The endocrine cells in the gastro–intestinal tract are referred to collectively as the enteric endocrine system.

Like all endocrine cells, cells in the enteric endocrine system secrete hormones in response to specific stimuli and they stop secreting the hormone when the stimuli are no longer present. In contrast to the endocrine glands in which essentially all cells produce hormones, the endocrine cells of the enteric endocrine system are dispersed over other types of epithelial cells in the mucosa of the stomach and small intestine.

Hormones are extremely important in controlling digestion secretions and most of them also affect the motility of some parts of the GIS, although the effect on the motility is less important than on the secretions. Two of them are cholecystokinin and secretin. Cholecystokinin stimulates the secretion of pancreatic enzymes and bile. It is produced in the mucosa of the duodenum and jejunum, mainly in response to the presence of breakdown products of fat, fatty acids, and monoglycerides in the intestinal contents. It has a potent effect in increasing contractility of the gallbladder, thus expelling the bile into the small intestine. At the same time, it inhibits stomach motility moderately, slowing the emptying of food from the stomach to give adequate time for digestion of fats in the upper intestinal tract. Secretin is secreted in the small intestinal epithelial cells of the duodenum, mainly in response to fatty acids and amino acids and stimulates secretion of fluids from the pancreas and liver. It has a mild effect in decreasing the motor activity of the stomach and therefore slows the emptying of gastric contents into the duodenum when the upper small intestine is already oversupplied with food products.

In addition to the hormones listed above, cells in the gastro-intestinal tract also secrete a wide range of other peptide regulators, such as: gastrin, ghrelin, motilin, gastric inhibitory polypeptide, etc.

3.4 Conclusion

In this chapter the elementary aspects of the GIS were briefly presented. Some basic concepts described in the previous chapter were introduced here from more global point of view, as well as the role of the principal actors in the complex digestive process.

Different sections of the digestive tube and its associated accessory organs have been described along the path of the ingested food, from the mouth to the anus. The structure of the intestinal wall has been discussed, and it was shown that the epithelium of the gut is not a monotonous sheet of functionally identical cells. In fact, as ingesta travels through the intestine, it is sequentially exposed to regions having epithelia of very different characteristics.

Five digestive functions have been identified and presented. The motility consists in the ability of moving spontaneously, independently, and in a coordinated manner the muscles composing the GIS. The secretion consists in adding to the luminal contents watery solutions containing ions, biliary components, and digestive enzymes. The digestion makes the food absorbable by the body by breaking it down and dissolving it into simpler chemical compounds. The absorption is the transport of water, electrolytes, and essentially all dietary organic molecules from the lumen across the epithelium into the body. The immuno-defense is vital to the good performance of the digestive system against potentially dangerous agents. The digestive processes of the principal nutrients have also been discussed.

Description of the control of the digestive functions through the enteric nervous system and the enteric endocrine system, modulated by the central nervous system, highlighted the complexity of the regulation of this system by means of a combination of nervous, mechanical, and hormonal signals.

Part II

Biophysical Modeling

The second part of this dissertation is dedicated to electrophysiological models of the bowel, its aspects related to discretization, and the extraction from the simulations of the parameters that can be measured in in vivo experiments for purposes of validation. In the first chapter of this part, three conceptually different approaches of intestine computer modeling are presented and discussed. Basic electrophysiological concepts are used to obtain mathematical descriptions of the smooth muscle tissue. The second chapter is an exploration of the temporal and spatial discretization effects, so as to inspect the numerical instabilities and errors affecting the results. The last chapter of this part presents some approaches permitting a comparison between simulated data and recorded real signals.

Chapter 4

Modeling Intestinal Electrophysiology

Computer models are a simplified representation of reality, that attempt to gain insight into a process or a system behaviour. The mathematical language is used to describe quantitatively the relationships among the entities in a system, and to demonstrate the effect that various actions will have on the global behaviour.

Computer models have become a useful aspect of the study of many natural systems in physics, chemistry, biology, and human systems in economics and social science. The main research topic, and the most important for biological and medical applications, is the development of realistic models of human organs. The ever–increasing computer power made possible to simulate models of full organs with a realistic size and a detailed description of the cellular electrophysiological properties. Among the advantage of models, we find of course the reproducibility of experiments, which is often difficult to achieve in real experiments. The access to data is another big advantage of *in silico* experiments, permitting to know the value of every variable of the model at any time and location. This is limited, in the case of real experiments, to some variables and moreover some regions remain inaccessible. Tissue preparation, including the alteration of tissues properties, is also easy and reliable, at the opposite of real experiments, for which tissue preparation is difficult and requires a careful validation. The drawback of models are the experiment duration, up to 1000 times slower than real time, and of course the relevance to the original real system, for which model validation is crucial.

The investigation of gut slow waves and spikes is a topic of intense research both from the experimental and simulation viewpoints so as to understand their functions and anomalies. Many experiments, in different ways, have brought information upon the electromechanical behaviour of small gastro–intestinal segments, but the motility of the gastro–intestinal tract is still not fully understood. Also, due to numerous technical problems, investigation of the spatial and temporal organization of the whole system is today possible only with the help of models [10].

Computer modeling of slow waves propagation in the intestinal tract has been conducted since the 1960s, when pioneering works had an essential impact on approaches used in later studies. The basic concept, i.e. simulating the electrical activity by an array of coupled relaxation oscillators, showed qualitative agreement with experimental observations. However, the pure mathematical approaches chosen to simulate the intrinsic electrical activity could not be rationalized in terms of the electrical behaviour of single cells.

To establish a correspondence between model parameters and morphological and electrophysiological data of a single cell, a sophisticated mathematical description is required. Ionic

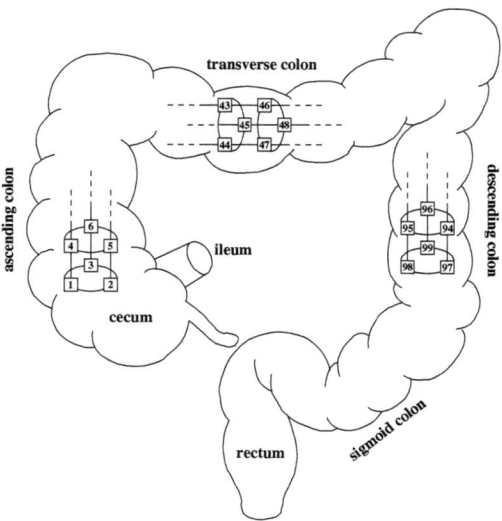

Figure 4.1: Arrangement of oscillators in the human colonic model for the simulation of the electrical control activity.

membrane models provide the most elaborate and realistic way to model action potential, but are unfortunately computationally expensive. Simplified reaction–diffusion models have been proved to be useful for gaining mathematical insight into the membrane nonlinearities underlying the electrical properties. Their parameters however have no direct physical interpretation.

The objective of this chapter is to investigate three different approaches to obtain a computer model simulating the electrical activity of the intestinal tissue. The first approach is based on a chain of relaxation oscillators. The second approach is founded on experimentally measurable biophysical variables, particularly attractive for the quality of its results, but quite demanding in terms of data processing. The last one uses a system description that belongs to the FitzHugh–Nagumo family, a combination of mathematical and physiological synthesis process.

4.1 Bardakjian Model

Bardakjian proposed a computer model of human colonic electrical activity using the recordings obtained from the colon of 17 patients undergoing cholecystectomy [11]–[12]. Recordings were obtained along the whole colon by direct implantation of three to six teflon coated bipolar stainless steel electrodes per subject [13]. Knowledge of the frequency content of the electrical activity permitted to synthesize a mathematical description of a population of 99 oscillators, which were used to simulate the *in vivo* colonic features. The oscillators are arranged in a tubular structure of 33 bidirectionally coupled rings, where each ring consists of three bidirectionally coupled oscillators as shown in figure 4.1.

The proximal nine rings represent the ascending colon, the next 12 the transverse colon, and the last 12 rings the descending and the sigmoid colon. The number of rings and the number of oscillators per ring were arbitrarily chosen, but were found to be adequate to reproduce the observed features of an intact colon.

Each oscillator corresponds to a small region of the colon, and reproduces the most relevant frequency components of its intrinsic electrical activity. The characteristics of individual oscillators could be obtained by dividing the colon into small segments, which is not actually feasible. Hence, the choice of the intrinsic model parameters representing the uncoupled state was governed mainly by an attempt to reproduce the observed *in vivo* features of the intact colon (coupled state).

The principal steps for obtaining the Bardakjian computer model of the colon are presented in the following of this section. At first, the generic equation of the isolated oscillator is derived from the second–order oscillator equations. Afterwards, oscillators are coupled together and disposed in a tubular structure as explained above and shown in figure 4.1.

4.1.1 The Generic Equation of the Isolated Oscillator

The pattern of the electrical activity in the colon is intricate. The most prominent feature is the simultaneous presence of multiple and variable frequency components at the same site, and has been well defined in the frequency domain. Linkens [14] simulated the colorectal myoelectrical activity of humans using a Van der Pol oscillator and a modified version of it but it appeared to be inadequate since it did not provide a direct control over the frequency content. Bardakjian also chose to describe the electrical behaviour of a small tissue region by synthesizing the differential equations of an oscillator whose solution is a truncated Fourier series. The parameters of an oscillator representing a colonic site would then be the coefficients of the truncated Fourier series of the control wave at that site when it is electrically isolated from its neighboring sites. Chua and Green [15] introduced this approach for the synthesis of nonlinear periodic systems.

Objectives

We want to define a system that fulfil the following properties:

1. The dynamics of the system has to possess a stable limit cycle.
2. One of the state variables has to generate, in the limit cycle, a given periodic function.
3. The equations of the system should be as simple as possible.

We will describe a simple system fulfilling the property 1. After having applied a suitable transform (diffeomorphism: invertible and differentiable), the system will satisfy the property 2.

A Simple System with Limit Cycle

Let us consider the variables μ_1 and μ_2 depending on time, $\dot{\mu}_1$ and $\dot{\mu}_2$ being the corresponding first derivatives. The second–order autonomous system $\hat{S}(\mu_1, \mu_2)$ is defined as follows:

$$\dot{\mu}_1 = \omega \left[\mu_2 + \mu_1 \left(1 - \mu_1^2 - \mu_2^2\right)\right], \quad (4.1)$$
$$\dot{\mu}_2 = \omega \left[-\mu_1 + \mu_2 \left(1 - \mu_1^2 - \mu_2^2\right)\right]. \quad (4.2)$$

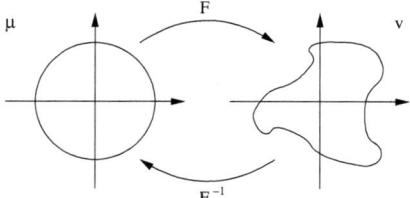

Figure 4.2: Direct transform $\mathcal{F}$ and inverse transform $\mathcal{F}^{-1}$, from the v space to the μ space.

Excluding the point $(\mu_1, \mu_2) = (0,0)$, which is the only unstable equilibrium point of $\hat{S}(\mu_1, \mu_2)$, equations 4.1 and 4.2 can be transformed using the polar coordinates $r^2 = \mu_1^2 + \mu_2^2$ and $\varphi = \arctan \frac{\mu_1}{\mu_2}$ into:

$$\dot{r} = r\left(1 - r^2\right), \qquad (4.3)$$
$$\dot{\varphi} = \omega. \qquad (4.4)$$

The equation 4.3 shows that $\dot{r} < 0$ for all $r > 1$, $\dot{r} > 0$ for all $0 < r < 1$ and that the only positive value of r for which $\dot{r} = 0$ is $r = 1$ (except of course for $r = 0$). Consequently, all solutions approach $r = 1$ as a limit, and at this point equations 4.1 and 4.2 become:

$$\dot{\mu}_1 = \omega \mu_2, \qquad (4.5)$$
$$\dot{\mu}_2 = -\omega \mu_1. \qquad (4.6)$$

In the limit, the system $\hat{S}(\mu_1, \mu_2)$ has a unique nonconstant periodic solution $\mu_1 = \sin(\omega t + \theta)$ and $\mu_2 = \cos(\omega t + \theta)$, where the constant phase angle θ depends only on the initial condition. Since the initial time t_0 is arbitrary, let us define $t_0' = t_0 - \left(\frac{\theta}{\omega}\right)$. For any initial condition $(\mu_1(t_0'), \mu_2(t_0'))$, the steady–state solution is $\mu_1 = \sin(\omega t)$ and $\mu_2 = \cos(\omega t)$. Thus, without loss of generality, it is implicitly assumed that $\theta = 0$.

Choice of the Transform

The generic solution of a system having a prescribed periodic stable behaviour is expressed as a Fourier series:

$$v_1(t) = \sum_{k=-\infty}^{\infty} \alpha_k e^{jk\omega t} \qquad (4.7)$$

where α_k, the Fourier coefficients of the function $v_1(t)$, have been obtained from a frequency analysis of experimental recordings. By the definition of Fourier series, $v_1(t)$ must obey the Dirichlet condition: $v_1 : \mathbb{R} \to \mathbb{R}$, must be periodic, bounded, piecewise–continuous, and have a finite number of extrema and discontinuities in one period.

We would like to obtain a transform $(\mu_1, \mu_2) \xrightarrow{\mathcal{F}} (v_1, v_2)$ such that $(\cos(\omega t), \sin(\omega t)) \longmapsto (v_1(t), v_2(t))$, where v_1 is given by equation 4.7. Figure 4.2 illustrates the direct transform $\mathcal{F}$ and its inverse for passing from v–space to μ–space.

Section 4.1 Bardakjian Model

An approximated form of equation 4.7 may be considered, so as to obtain a Fourier series of order K, which is represented with the following finite sum:

$$v_1(t) = a_0 + \sum_{k=1}^{K} [a_k \cos(k\omega t) + b_k \sin(k\omega t)]. \qquad (4.8)$$

Parameters a_k, b_k, K, and ω are determined, so as to best reproduce the actual observations. The synthesis procedure is directed towards generating equation 4.8. One also first generates waveforms $\sin(\omega t)$ and $\cos(\omega t)$, which are combined to produce the periodic solution of equation 4.8.

Any periodic waveform expressed as a Fourier series may be written as a polynomial of sine and cosine functions at the fundamental frequency. In particular, the kth Chebyshev polynomials of the first and second kind, T_k and U_k respectively, are defined as:

$$\cos(k\alpha) = T_k(\cos(\alpha)), \qquad (4.9)$$
$$\sin(k\alpha) = \sin(\alpha) U_{k-1}(\cos(\alpha)), \qquad (4.10)$$

and can be computed using the relations:

$$T_k(x) = \frac{k}{2} \sum_{j=0}^{\frac{k}{2}} (-1)^j \frac{(k-j-1)!}{j!(k-2j)!} (2x)^{k-2j}, \qquad (4.11)$$

$$U_k(x) = \sum_{j=0}^{\frac{k}{2}} (-1)^j \frac{(k-j)!}{j!(k-2j)!} (2x)^{k-2j}. \qquad (4.12)$$

Now, using equations 4.5 and 4.6 and equations 4.9 and 4.10, the generic equation of an isolated oscillator 4.8 becomes:

$$v_1(t) = a_0 + \sum_{k=1}^{K} [a_k T_k(\sin(\omega t)) + b_k \cos(\omega t) U_{k-1}(\sin(\omega t))]. \qquad (4.13)$$

In a similar way, high order differential equations having equation 4.8 as unique steady-state solution can be developed.

Bardakjian preferred to consider a simplified form of equation 4.8. He described the output of an electrically isolated oscillator representing a small region of the colon supposing $b_k = 0$ for $k > 1$ and obtaining the following transformation:

$$v_1(t) = a_0 + b_1 \sin(\omega t) + \sum_{k=1}^{K} a_k \cos(k\omega t). \qquad (4.14)$$

Equation 4.13 suggests the following expression for the transform $\mathcal{F}$:

$$v_1(\mu_1, \mu_2) = a_0 + b_1 \mu_1 + \sum_{k=1}^{K} a_k T_k(\mu_2), \qquad (4.15)$$

$$v_2(\mu_1, \mu_2) = \mu_2. \qquad (4.16)$$

This transformation $\mathcal{F}$ fulfils the three properties mentioned above, and its inverse $\mathcal{F}^{-1}$ is given by:

$$\mu_1(v_1, v_2) = \frac{1}{b_1} \left[v_1 - a_0 - \sum_{k=1}^{K} a_k T_k(v_2) \right] \triangleq \rho(v_1, v_2), \qquad (4.17)$$

$$\mu_2(v_1, v_2) = v_2. \qquad (4.18)$$

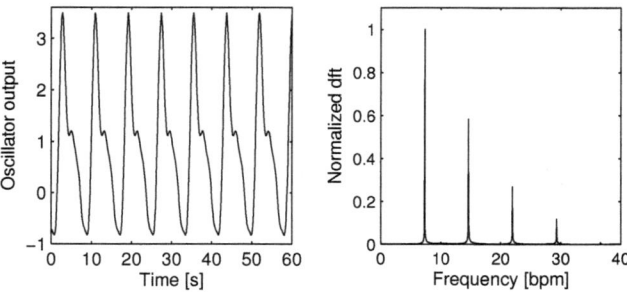

Figure 4.3: Output of the isolated oscillator. Parameters used in this simulation are $f = 7.5$, $a_0 = a_1 = b_1 = 1$, $a_2 = 0.8$, $a_3 = 0.4$, and $a_4 = 0.2$.

Derivation of the Equation for an Isolated Oscillator

Then using equations 4.1 and 4.2 and equations 4.15 and 4.16, the differential equations governing the generic isolated oscillator are:

$$\dot{v}_1 = \omega \left\{ b_1 \left[v_2 + \rho \left(1 - \rho^2 - v_2^2 \right) \right] \right.$$
$$+ \left[-\rho + v_2 \left(1 - \rho^2 - v_2^2 \right) \right]$$
$$\left. \cdot \left[\sum_{k=1}^{K} a_k \frac{\mathrm{d}T_k\left(v_2\right)}{\mathrm{d}v_2} \right] \right\}, \tag{4.19}$$

$$\dot{v}_2 = \omega \left[-\rho + v_2 (1 - \rho^2 - v_2^2) \right], \tag{4.20}$$

The output of an isolated oscillator is shown on the left panel of figure 4.3. The uncoupled state is characterized by a periodic repetition of the waveform defined by the oscillator parameters b_1 and $a_i, i = 0, \ldots, K$ at a specific frequency f. Performing a frequency analysis on the output of the isolated oscillator, the frequency $f_{20} = 7.5$ bpm and its three first harmonics are obtained.

4.1.2 The Generic Equation for a Population of Bidirectionally Coupled Oscillators

In the intact colon each small region representing an oscillator is coupled to its adjacent oscillators through intercellular connections. To simulate an intact colon, let the nth oscillator be coupled to its adjacent oscillators in a bidirectional fashion and define a type of coupling such that equations 4.3 and 4.4 are modified to:

$$\dot{r}_n = r_n \left(1 - r_n^2 \right), \tag{4.21}$$
$$\dot{\varphi}_n = \omega_n (1 + \kappa_n), \tag{4.22}$$

where

$$\kappa_n = \sum_{m \in I_{c,n}} c_{mn} \left(v_{1,m} - v_{1,n} \right). \tag{4.23}$$

Table 4.1: Intrisic oscillator-parameters of the Bardakjian model. $a_{0,n} = a_{1,n} = b_{1,n} = 1$ for $n = 1\ldots 99$, c_{circ} is the coupling between adjacent cells of the same ring, and c_{long} is the coupling between adjacent cells of neighboring rings.

Ring	Oscillator numbers			Intrinsic frequency [bpm]			Fourier coefficients			Coupling	
	i	j	k	f_i	f_j	f_k	$a_{2,i/j/k}$	$a_{3,i/j/k}$	$a_{4,i/j/k}$	c_{circ}	c_{long}
1	1	2	3	2.75	2.5	3	2	0.8	0.4	0.1	0.1
2	4	5	6	3	2.75	3.5	0.8	2	0.4	0.101	0.101
3	7	8	9	4	3.5	4.5	2	0.8	0.4	0.102	0.102
4	10	11	12	5	4.5	5.5	2	0.8	0.4	0.103	0.103
5	13	14	15	6	5.5	6.5	0.8	0.4	0.2	0.104	0.104
6	16	17	18	7	6.5	7.5	0.8	0.4	0.2	0.105	0.105
7	19	20	21	8	7.5	8.5	0.8	0.4	0.2	0.106	0.106
8	22	23	24	9	8.5	9.5	0.8	0.4	0.2	0.108	0.108
9	25	26	27	10	9.5	10.5	0.2	0.8	0	0.11	0.11
10	28	29	30	10.2	10.4	10.6	0.8	0.2	0	0.112	0.112
11	31	32	33	10.4	10.2	10.8	0.8	0.2	0	0.114	0.114
12	34	35	36	10.6	10.8	11	0.8	0.2	0	0.116	0.116
13	37	38	39	10.8	10.6	11.2	0.8	0.2	0	0.118	0.118
14	40	41	42	11	11.2	11.4	0.8	0.4	0	0.12	0.12
15	43	44	45	11.2	11.4	10.8	0.2	0.8	0	0.125	0.125
16	46	47	48	11.4	11.2	11	0.8	0.4	0	0.13	0.13
17	49	50	51	11.3	11	10.8	0.8	0.4	0	0.135	0.135
18	52	53	54	11.4	11.2	11	0.2	0.8	0	0.14	0.14
19	55	56	57	11.2	11.4	11.3	0.8	0.4	0	0.145	0.145
20	58	59	61	11	11.2	11.4	0.8	0.4	0	0.15	0.15
21	61	62	63	10.8	10.6	11.2	0.2	0.8	0	0.155	0.155
22	64	65	66	10.6	10.8	11	0.8	0.4	0	0.16	0.16
23	67	68	69	10.4	10.2	10.8	0.8	0.2	0	0.165	0.165
24	70	71	72	10.2	10.4	10.6	0.8	0.2	0	0.17	0.17
25	73	74	75	10	9.5	10.5	0.8	0.2	0	0.172	0.172
26	76	77	78	9	8.5	9.5	0.8	0.2	0	0.174	0.174
27	79	80	81	8	7.5	8.5	0.8	0.2	0	0.176	0.176
28	82	83	84	7	6.5	7.6	0.8	0.2	0	0.178	0.178
29	85	86	87	6	5.5	6.5	0.8	0.2	0	0.18	0.18
30	88	89	90	5	4.5	5.5	2	0.4	0	0.182	0.182
31	91	92	93	4	3.5	4.5	2	0.4	0	0.184	0.184
32	94	95	96	3	2.75	3.5	0.4	2	0.2	0.186	0.186
33	97	98	99	2.75	2.5	3	0.2	0.4	2	0.188	0

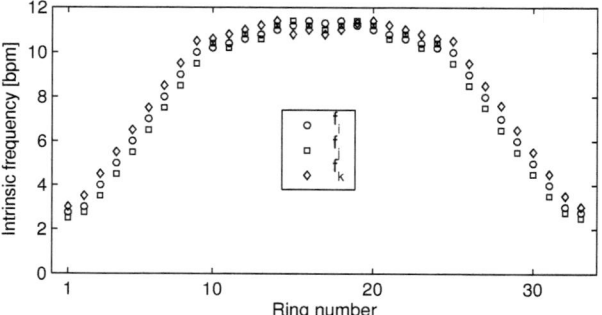

Figure 4.4: Intrinsic frequency distribution in the longitudinal direction of Bardakjian model. At each ring number corresponds three oscillator frequencies (f_i, f_j, f_k) in beat per minute, which are displayed with $\bigcirc$, $\square$, and $\diamond$, respectively.

$I_{c,n}$ is a coupling index set representing the input to the nth oscillator from its adjacent oscillators and $c_{m,n}$ is the coupling factor representing the portion of the output of the mth oscillator to be fed to the nth oscillator. $v_{1,m}$ and $v_{1,n}$ are the outputs of the mth and nth coupled oscillators, respectively.

The equation of the nth coupled oscillator can be obtained by defining the second order coupled system $\hat{S}_n(\mu_{1,n}, \mu_{2,n})$:

$$\dot{\mu}_{1,n} = \omega_n \left[\mu_{2,n}(1 + \kappa_n) + \mu_{1,n}(1 - \mu_{1,n}^2 - \mu_{2,n}^2) \right], \qquad (4.24)$$

$$\dot{\mu}_{2,n} = \omega_n \left[-\mu_{1,n}(1 + \kappa_n) + \mu_{2,n}(1 - \mu_{1,n}^2 - \mu_{2,n}^2) \right]. \qquad (4.25)$$

The generic equation for a population of coupled oscillators is then synthesized as follows:

$$\dot{v}_{1,n} = \omega_n \Big\{ b_{1,n} \left[v_{2,n}(1 + \kappa_n) + \rho_n(1 - \rho_n^2 - v_{2,n}^2) \right]$$
$$+ \left[-\rho_n(1 + \kappa_n) + v_{2,n}(1 - \rho_n^2 - v_{2,n}^2) \right]$$
$$\cdot \left[\sum_{k=1}^{K} a_{k,n} \frac{\mathrm{d}T_k(v_{2,n})}{\mathrm{d}v_{2,n}} \right] \Big\} \qquad (4.26)$$

$$\dot{v}_{2,n} = \omega_n \left[-\rho_n(1 + \kappa_n) + v_{2,n}(1 - \rho_n^2 - v_{2,n}^2) \right] \qquad (4.27)$$

where

$$\rho_n = \frac{1}{b_{1,n}} \left[v_{1,n} - a_{0,n} - \sum_{k=1}^{K} a_{k,n} T_k(v_{2,n}) \right]. \qquad (4.28)$$

The parameters and the coupling factors corresponding to the 99 oscillators of the model are given in the table 4.1, whereas the measured intrinsic frequency distribution of the colon model is depicted in figure 4.4. The measured gradient in the longitudinal direction is symmetric and characterized by an increase in the ascending section, a plateau in the transverse, and a decrease in the descending colon. No significant difference has been observed in the circumferential

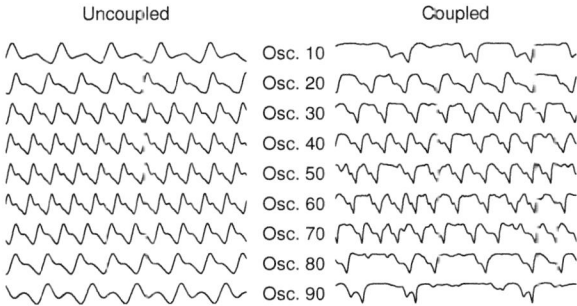

Figure 4.5: Representation of the outputs of nine oscillators. The uncoupled case, at the left, and the coupled case, at the right, are displayed for a time window of one minute.

direction. Thus the intrinsic frequencies of the three oscillators of each ring in the model are practically equal, as shown in figure 4.4.

The outputs of nine oscillators representing a few sites in the proximal, middle, and distal colon are shown in figure 4.5. Uncoupled and coupled states are presented, in order to observe the changes in the outputs. The oscillators in the uncoupled state have a regular and stable rhythm, while in the coupled state, they become variable in frequency and amplitude. The unsteady comportment in the coupled configuration is the expression of the interactions with the neighbor oscillators.

4.2 Miftakhov Model

The Nobel prize winning work of Hodking and Huxley on the excitable properties of the giant squid axon paved the way to realistic membrane models [16]. The equations are expressed in terms of membrane capacitances and conductances. They have been modified by several scientists to model the autorhythmic behaviour of cardiac muscles [17], while Linkens [18] suggested modifications for the smooth muscle cells. Using an electronic implementation of the model, it has been shown that the model can reproduce slow waves as well as spike activity [19]. The concepts of membrane capacitance and ionic conductance, of importance to physiologists, are particularly relevant for the current research on the properties of smooth muscle cells in the digestive tract.

Generic Membrane Models

A ionic membrane model typically includes three components connected in parallel as depicted in figure 4.6. The cell membrane consists of a dielectric lipid bilayer that exhibits capacitive properties. The enclosed fluid is called the intracellular medium, and the interstitial fluid is called the extracellular medium.

The electric charge $Q_m = C_m V_m$ accumulated at the membrane surface (per unit area of membrane) of an isolated cell is proportional to the transmembrane potential V_m, with C_m

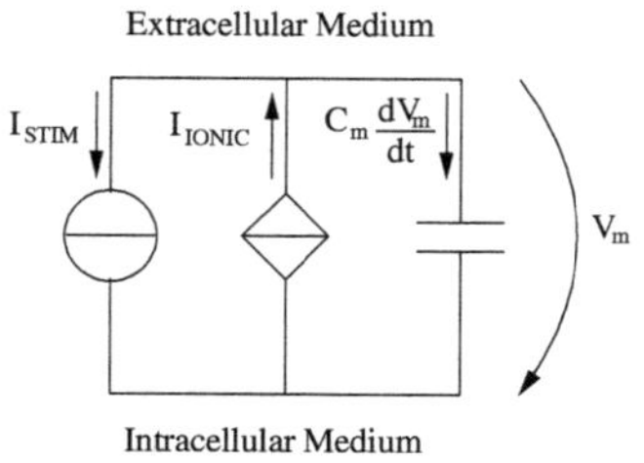

Figure 4.6: Equivalent current–source model for the cardiac membrane.

being the transmembrane capacitance (per unit area of membrane). The membrane contains ion channels, pumps, and exchangers that permit ion exchanges between intracellular space and extracellular space. The model described in this section employs only the ionic channels for producing or transducing ionic fluxes through the cellular membrane, that is, the phenomena associated with pumps and exchangers are not included. The flows of ions through the membrane generate ionic currents, I_{ionic}, which depend on the value of the electrochemical gradient between the two media. Then, the total transmembrane currents influence V_m, which evolves in time according to the following equation:

$$C_m \frac{dV_m}{dt} = -\sum I_{\text{ionic}} + I_{\text{stim}}, \qquad (4.29)$$

where I_{stim} is the external stimulus current that electrically stimulates the cellular membrane, as is illustrated in figure 4.6.

Ionic channels are the elementary components of the cellular membrane, capable of producing or transducing ionic fluxes through the cellular membrane. The structure of a channel is modeled by a sequence of independent elements, each of which is associated to a specific function [20]–[21]. If only a restricted set of ion types is allowed to cross the channel, the latter acts as a selective filter. A gate is a channel that randomly opens and closes [2]. When it depends on an inactivation mechanism, the channel behaves as a receptor. The concatenation of these three primitive functions (filter, gate, and receptor) allows us to emulate the complex behaviour of a single channel. Through each channel type flows a specific ionic current. The charge of accumulated ionic currents appears in equation 4.29, while the dynamics of a single current can be expressed using Ohm's law as follows:

$$I_k = g_k\left(V_m - E_k\right) \qquad \text{and} \qquad g_k = \tilde{g}_k y \qquad (4.30)$$

where I_k is the current associated to ion k, g_k is the conductance, $\tilde{g}_k$ is the maximal conductance of the corresponding ionic channel, and y is called the gating variable expressing the mean state of these ion channels. E_k is the equilibrium or Nernst potential [3]–[20] of ion k and depends on intracellular and extracellular concentrations and the valence of the same ion (see chapter 1 of the dissertation). The equilibrium is reached when the electrical and the chemical gradients are equal and inverse, with respect to all ions. The sign of the electrochemical gradient determines also the direction of movement of the ionic current across the membrane. An inward current is defined as the movement of a positive charge into a cell, or a negative charge out of a cell. It is considered negative by convention and causes depolarisation. An outward current is defined as the movement of a positive charge out of a cell and causes the membrane potential to become more negative (repolarisation)

In the simplest two–state model of a gate, the variable y can be either in an open state ($y = 1$) or in a closed state ($y = 0$). The transition of the gating variable y can be written as:

$$\text{Closed} \underset{\beta_y}{\overset{\alpha_y}{\rightleftharpoons}} \text{Open}. \quad (4.31)$$

The gate can switch from the open to the closed state with transition rate α_y, and, inversely, from the closed to the open state with a rate β_y. Statistical signal processing provides appropriate tools for modeling random transitions between a few discrete levels. In particular, using hidden Markov models, the fraction $y(t)$ of open gates at time t is expressed as follows:

$$\frac{dy}{dt} = \overbrace{\alpha_y (1 - y)}^{C \to O} - \overbrace{\beta_y y}^{C \leftarrow O} \quad (4.32)$$

where the first element of the right hand side equation 4.32 expresses the part of closed gates switching to open state, while the second expresses the part of open gates reaching the closed state. The difference between these two elements gives the variation of the open–gate ratio. The gate rates α_y and β_y are voltage–dependent. Under steady conditions V_m is constant and thus $dy/dt = 0$, and the solution to equation 4.32 can be characterized by its asymptotic value y_∞ and its time constant τ_y, such as:

$$\tau_y = \frac{1}{\alpha_y + \beta_y} \quad \text{and} \quad y_\infty = \frac{\alpha_y}{\alpha_y + \beta_y}. \quad (4.33)$$

Then, equation 4.32 can be written in another form:

$$\frac{dy}{dt} = \frac{y_\infty - y}{\tau_y}. \quad (4.34)$$

Considering that the system reaches instantaneously the steady–state, the function y_∞ describes the evolution of the gating variables with respect to the transmembrane potential V_m. The characteristics of the response velocity in opening or closing the channel is given by τ_y. Since τ_y depends on α_y and β_y, the relaxation time of a channel is also modulated by V_m.

Formulation of the Miftakhov Model

The ionic model proposed by Miftakhov [22]–[23] incorporates the main electrophysiological and morphological characteristics of the smooth muscle cell. The schematic representation and its equivalent circuit, representing the basic equations, are illustrated in figure 4.7. The model includes five ionic currents: I_{Ca}^f and I_{Ca}^s are the fast and slow inward calcium currents via T–type[1] and L–type[2] voltage–dependent Ca^{2+} channels. I_K and I_{Ca-K} are outward potassium currents, to which associated channels are respectively calcium–activated and voltage–dependent.

---

[1] Widely distributed in excitable and non–excitable cells, inactivate rapidly, and allows oscillation of membrane potential.
[2] Structurally similar to T–type channels, large sustained conductance, inactivate slowly, and are responsible for plateau phase of action potential.

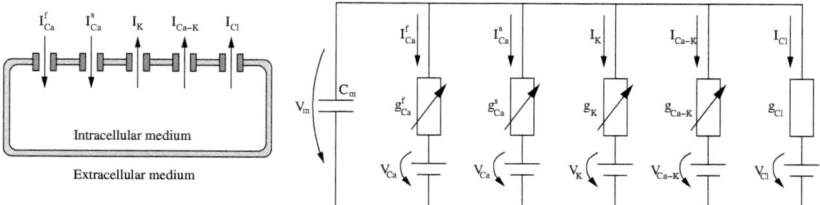

Figure 4.7: On the left, schematic representation of the main ion flows involved in the membrane exchanges, as well as the outward or inward direction of their driving force, for the Miftakhov model. Its equivalent current–source circuit is displayed on the right.

I_{Cl} is the leak chloride current. The sum of the five currents can be inserted into equation 4.29, which becomes:

$$C_m \frac{dV_m}{dt} = -\left(I_{Ca}^f + I_{Ca}^s + I_K + I_{Ca-K} + I_{Cl}\right) + I_{stim} \tag{4.35}$$

The right side of figure 4.7 represents the relation existing between the transmembrane current and the evolution of the transmembrane potential using a simple electric scheme. Ionic currents are defined as in equation 4.30, and can be listed as:

$$I_{Ca}^f = g_{Ca}^f \left(V_m - V_{Ca}\right), \tag{4.36}$$

$$I_{Ca}^s = g_{Ca}^s \left(V_m - V_{Ca}\right), \tag{4.37}$$

$$I_K = g_K \left(V_m - V_K\right), \tag{4.38}$$

$$I_{Ca-K} = g_{Ca-K} \left(V_m - V_{Ca-K}\right), \tag{4.39}$$

$$I_{Cl} = g_{Cl} \left(V_m - V_{Cl}\right). \tag{4.40}$$

Each current depends on the conductance and on the constant reversal potential associated to the specific channel. The conductance can be either constant or modulated by specific parameters. Variation of its value is regulated by criteria of different natures, in particular voltage and ion concentration. The expression of the conductances related to the model are:

$$g_{Ca}^f = \tilde{g}_{Ca}^f m^3 h, \tag{4.41}$$

$$g_{Ca}^s = \tilde{g}_{Ca}^s x, \tag{4.42}$$

$$g_K = \tilde{g}_K n^4, \tag{4.43}$$

$$g_{Ca-K} = \tilde{g}_{Ca-K} [Ca^{2+}] / \left(0.5 + [Ca^{2+}]\right), \tag{4.44}$$

$$g_{Cl} = \tilde{g}_{Cl}, \tag{4.45}$$

where $\tilde{g}_{Ca}^f$, $\tilde{g}_{Ca}^s$, $\tilde{g}_K$, $\tilde{g}_{Ca-K}$, and $\tilde{g}_{Cl}$ are the maximal conductances of the associated channels, while m, h, x, and n are the gating variables defining the characteristics of the channels.

Statistical independent gate models can be combined in series to obtain complex gating models, while keeping the number of independent parameters moderate. Since the gates are placed in series, the ionic flow is present only if all the gates are open. The diagram of the kinetic states is also characterized by one open state and various closed states. In particular, equation 4.43 indicates that four identical gating variables n are combined to describe the characteristics of

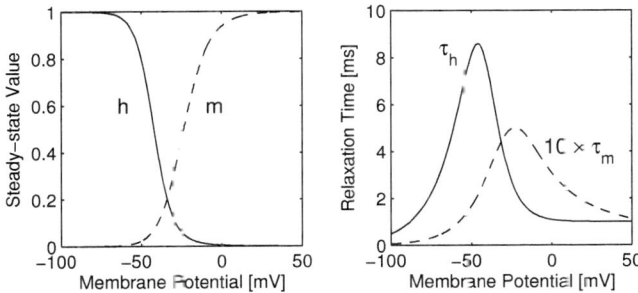

Figure 4.8: Representations of the parameters characterising the gating kinetics of the inward Ca^{2+} current. On the left side, the steady-state curves, and on the right, the relaxation time for the gating variables of the channel are displayed.

the resulting gate. The gate is open if and only if all the four gates in the channel are open. The probability of the gate to be open is thus the product of the gating variables, and is also of the form n^4.

Some ionic currents are governed by antagonistic gating variables, one of which opening at high V_m and the other closing in the same condition. In equation 4.41, m and h are classical examples of antagonist gate variables. Left part of figure 4.8 shows the sigmoidal dependences of these two variables on V_m, typical of certain gating variables. Under these conditions, ionic flow is present because of the different relaxation times of the antagonist gates. The evolution of the relaxation time τ of the two variables, with respect to V_m, is depicted on the right side of figure 4.8. The gate associated to the variable m reacts much faster than that described by h.

Equation 4.44 illustrates the case of a conductance modulated by an ionic concentration, in this case of the calcium. The simplest expression of the conductance is given by equation 4.45. The chloride current is related linearly to the transmembrane potential, and its value depends on the deviation from the equilibrium value (Ohm's law).

The calcium concentration $[Ca^{2+}]$ and the gating variables m, h, x, and n necessary for the evaluation of the ionic currents in the cellular model are defined by:

$$m = \frac{\alpha_m}{\alpha_m + \beta_m}, \tag{4.46}$$

$$\frac{dh}{dt} = \frac{\alpha_h(1-h) - \beta_h h}{\alpha\lambda}, \tag{4.47}$$

$$\frac{dn}{dt} = \frac{\alpha_n(1-n) - \beta_n n}{\alpha\lambda}, \tag{4.48}$$

$$\frac{dx}{dt} = \frac{1 + \left[e^{0.15(-V_m - 50)}\right]^{-1} - x}{\alpha\tau_x}, \tag{4.49}$$

$$\frac{d[Ca^{2+}]}{dt} = \frac{\rho}{\alpha}\left[K_c x(V_{Ca} - V_m) - [Ca^{2+}]\right], \tag{4.50}$$

where ρ and K_c are parameters related to the dynamics of the calcium channel, α is a numerical constant, λ is Plant factor, and τ_x is a time constant associated to the gating variable x. $\alpha_i, i =$

Table 4.2: The standard parameters and constants of the Miftakhov model.

Parameter	Value
C_m	2.5 μF cm^{-2}
$\tilde{g}_{Ca}^f$	0.51 mS cm^{-2}
$\tilde{g}_{Ca}^s$	0.004 mS cm^{-2}
$\tilde{g}_K$	0.3 mS cm^{-2}
$\tilde{g}_{Ca-K}$	0.03 mS cm^{-2}
$\tilde{g}_{Cl}$	0.003 mS cm^{-2}
φ_{rest}	-55 mV
φ_{Ca}	80 mV
φ_K	-75 mV
φ_{Cl}	40 mV
$[Ca^{2+}]$	10^{-4} mM
α	0.12
λ	12.5
τ_x	500 ms
K_c	$4.25 \cdot 10^{-3}$ mV^{-1}

m, h, x, n are the rates at which the gate m, h, x, or n switch from a closed to an open state and $\beta_i, i = m, h, x, n$ is the rate for the inverse. They depend only on the membrane potential and are given by the expressions:

$$\alpha_m = \frac{0.1\left(50 - V_m'\right)}{\exp\frac{50-V_m'}{10} - 1}, \tag{4.51}$$

$$\beta_m = 4\exp\frac{25 - V_m'}{18}, \tag{4.52}$$

$$\alpha_h = 0.07\exp\frac{25 - V_m'}{20}, \tag{4.53}$$

$$\beta_h = \left(1 + \exp\frac{55 - V_m'}{10}\right)^{-1}, \tag{4.54}$$

$$\alpha_n = \frac{0.01\left(55 - V_m'\right)}{\exp\frac{55-V_m'}{10} - 1}, \tag{4.55}$$

$$\beta_n = 0.125\exp\frac{45 - V_m'}{80}, \tag{4.56}$$

where $V_m' = (127 V_m + 8265)/105$.

Implementation

The model described above includes 15 parameters and constants [23] whose values are presented in table 4.2, and six variables: one for the membrane potential, four gating variables, and one for the calcium concentration. The complete Miftakhov model [22] includes 48 parameters and constants, and 16 variables for the description of a functional unit of the small intestine. Considering that the myoelectrical activity of each intestinal segment is under the control of the enteric

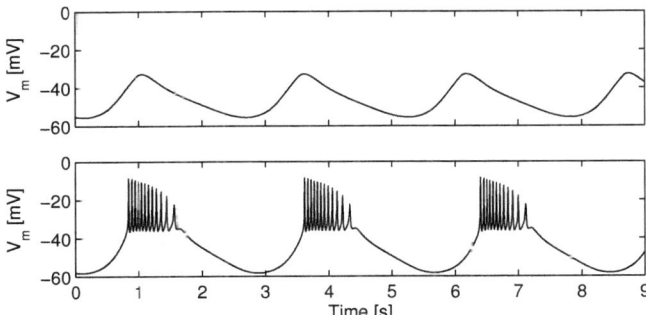

Figure 4.9: Two different situations of the membrane potential obtained using the Miftakhov model. Slow waves are obtained above using $\tilde{g}_{Ca}^f = 0.51$ mS/cm^2, while slow waves associated to spike activity are represented below using $\tilde{g}_{Ca}^f = 0.71$.

nervous system, which is functionally identified as a cholinergic[3] neurone, Miftakhov assumed that a functional unit of the small intestine is composed of a smooth muscle syncytium[4] and a cholinergic neurone. The cholinergic neurone generates high amplitude spikes that propagate along the unmyelinated axon to reach a cholinergic synapse, which is located on the smooth muscle membrane and modulates the properties of L-type voltage-dependent Ca^{2+} channels.

The extended version of the Miftakhov model is very expensive from a computational point of view, since it also demands a spatial discretization within the cell. Cellular models of the heart already reached larger complexity, up to the order of hundreds of thousand cells, for the stimulation of cardiac arrhythmia, for example. However, the time scale of the two simulated phenomena is sensibly different. Atrial fibrillation exhibits a frequency of about 400 bpm, and 20 s of real time simulation is sufficiently long to observe the process. The intestinal rhythms are between 1 and 20 bpm and some hours of simulation length is needed to obtain sufficient information. There is thus a computational time factor between 20 and 400. This is the major reason why it is today employed only for pharmacological validations, dealing with a single functional unit [24]-[25].

The slow oscillations of the membrane potential (slow waves), showed above in figure 4.9, are obtained using the simplified Miftakhov model. They have a frequency of 24 bpm and a constant amplitude of about 22 5 mV. The depolarization wave reaches its maximum in 1.5 s, followed by a short plateau and a repolarization phase to the resting value. A rise of $\tilde{g}_{Ca}^f$ from 0.51 to 0.71 mS/cm^2 causes significant qualitative and quantitative changes in the dynamics of the functional unit behaviour. The regular pattern of slow waves converts into bursting discharges. The frequency of the slow waves decreases a little, while its amplitude remains unchanged. Fast action potentials are generated at the top of the plateau with a frequency of about 15 Hz. Their firing level is high at the beginning, namely 20 mV, and decreases towards the end of the burst, to about 10 mV. In his paper [22], Miftakhov analysed and showed different electrical activity

---

[3]Liberating, activated by, or involving acetylcholine
[4]A multinucleate mass of protoplasm resulting from fusion of cells

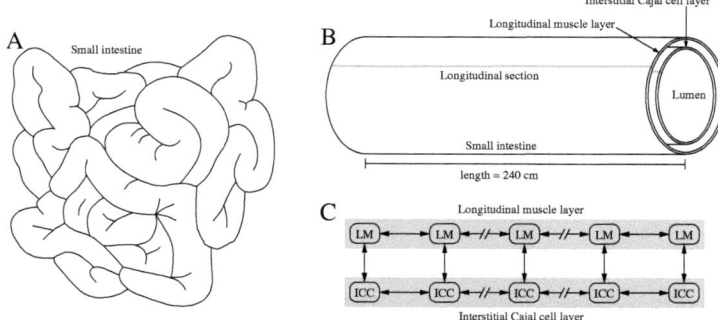

Figure 4.10: The human small intestine (A) is reduced to a 240 cm–long tubular structure (B) composed of two layers representing the longitudinal muscular cells and the interstitial Cajal cells. A cable is selected along the longitudinal direction of the pipe, and modeled using two coupled linear arrays (C). Both linear arrays are composed of bidirectionally coupled FitzHugh–Nagumo cells.

patterns of the functional unit with slight modifications of some of the parameters.

4.3 Aliev Model

Reaction–diffusion systems [26] correspond to the class of partial differential equations where the right hand side time–dependent can be divided into a reaction part and a diffusion part. These equations are an abstract model for pattern formation but in many cases have direct applications in the field of developmental biology, chemistry, optics, and other branches of applied mathematics. For biological and chemical systems, for example, reaction–diffusion equations represent a reduced description of a complicated set of reactions. The important aspect is that a simple combination of reaction, and transport due to the diffusion, is sufficient to produce a large variety of interesting patterns.

The computer model of the small intestine proposed by Aliev permits to investigate the propagation of slow waves in the whole small–intestinal tract [27]. It is based on a structure composed of two bidirectionally coupled layers, each of which respectively simulates the longitudinal muscle (LM) layer and the interstitial Cajal cell (ICC) layer (see figure 4.10). Both layers consist of a linear array of bidirectionally coupled cells, which are close in definition to the FHN model [28]–[29]. Each pair of ICC–LM cells represents a small patch of the intestine, and is connected to its two neighbor pairs.

The FHN model is one of the most famous reaction–diffusion models for the conduction of electrical impulses along a nerve fibre. The original heuristic model for the flow of electric current through the membrane of a giant squid axon was proposed by Hodgkin and Huxley [16]. The proposed four component model comprises the equations for the membrane current density, the sodium activation, the sodium inactivation, and the potassium activation in the nerve axon. FitzHugh and Nagumo simplified these equations using phase plane projections, and showed that

Table 4.3: Parameters of the two layers for the Aliev model.

Parameter	LM layer	ICC layer
κ	10	7
a	0.06	0.5
β	0	0.5
γ	8	8
ε	0.15	$\varepsilon_{\text{ICC}}(i)$
D_1	0.4	0.04
D_2	0.3	0.3

the dynamics of this model were qualitatively the same as the more complicated four–variable Hodgkin–Huxely model.

Formulation

A wide variety of FHN models have been used to simulate various biological systems, and in particular the electrical activity of the cardiac tissue. FHN models for the intestine present some differences compared to the standard cardiac or nervous formulations. The model proposed by Aliev is similar to the FHN model: its reaction–diffusion equations use two dimensionless state variables to represent the dynamics in each layer, and have, assuming no flux boundary condition ($\frac{\partial u}{\partial x} = 0$ at the extremities), the following form:

$$\frac{\partial u_{\text{LM}}}{\partial t} = \overbrace{\kappa u_{\text{LM}} (u_{\text{LM}} - a)(1 - u_{\text{LM}}) - v_{\text{LM}}}^{\text{reaction}} + \overbrace{D_1 \nabla^2 u_{\text{LM}} + D_2 (u_{\text{LM}} - u_{\text{ICC}})}^{\text{diffusion}}, \quad (4.57)$$

$$\frac{\partial v_{\text{LM}}}{\partial t} = \varepsilon (\gamma (u_{\text{LM}} - \beta) - v_{\text{LM}}), \quad (4.58)$$

$$\frac{\partial u_{\text{ICC}}}{\partial t} = \kappa u_{\text{ICC}} (u_{\text{ICC}} - a)(1 - u_{\text{ICC}}) - v_{\text{ICC}} + D_1 \nabla^2 u_{\text{ICC}} + D_2 (u_{\text{ICC}} - u_{\text{LM}}), \quad (4.59)$$

$$\frac{\partial v_{\text{ICC}}}{\partial t} = \varepsilon (\gamma (u_{\text{ICC}} - \beta) - v_{\text{ICC}}). \quad (4.60)$$

The variables u and v stand for the transmembrane potential and slow current, respectively. The indices LM and ICC specify the corresponding layer. The coefficient D_1 defines the coupling strength between cells of the same layer and D_2 between cells of different layers. The five model–specific constants κ, a, β, γ, and ε are "membrane" parameters that define the shape of the action potential pulse. Mathematically, these parameters control the number and type of fixed points, the reaction kinetics and the stability of the solutions of the reaction–diffusion equations. Their values, showed in table 4.3, are chosen in order to obtain an excitatory state for the LM layer and an oscillatory state for the ICC one.

Limit Cycle

The stationary homogeneous solutions of the reaction–diffusion system described by equations 4.57–4.60 are found by setting the time–derivative equal to zero and looking for intersections

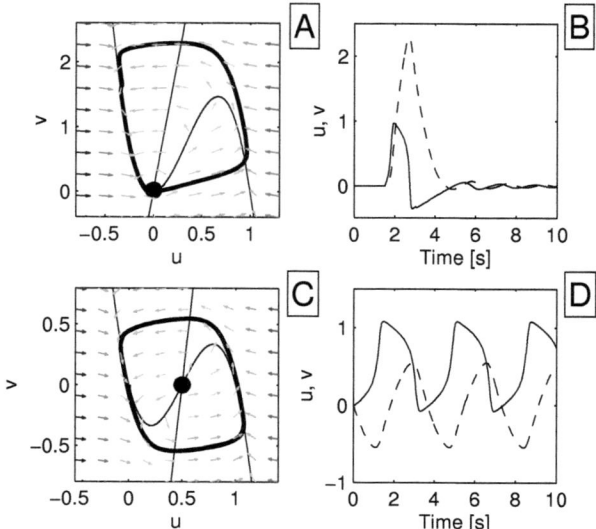

Figure 4.11: Phase space (u, v) and corresponding pulse solution for the case of LM layer (A and B) and ICC layer (C and D). Solutions of the two layer equations are displayed in the respective phase spaces (A and C) as bold lines. The nullclines, the equilibrium points, and the vector fields associated to the equation systems of the two layers are also illustrated. The solid line in C and D correspond to the u field, and the dashed one to v.

of the nullclines:

$$\kappa u (u - a)(1 - u) - v = 0, \qquad (4.61)$$
$$\varepsilon (\gamma (u - \beta) - v) = 0. \qquad (4.62)$$

When the nullclines intersect at a single point lying on an outer branch of the cubic nullcline and ε is small enough, the dynamics are excitable. This configuration is typically used in nervous and cardiac models, where the conduction of electrical impulses is simulated along nerve fibres or cardiac tissues. The single fixed point is stable, but perturbations over the excitation threshold cause a large excursion in phase space before return to the fixed point. When the nullclines intersect at a single point on the middle branch of the cubic nullcline and ε is small enough, the homogeneous state loses stability and uniform oscillations set in.

The two types of cells in the Aliev model, LM and ICC, generate the two basic cases discussed above and are displayed in figure 4.11. The phase space (u, v) of the LM layer is showed in figure 4.11(A) and presents a single stable fixed point at the origin, where the nullclines intersect. The intersection point on the cubic nullcline is located on its outer branch, determining an excitable dynamic. Thus, the system remains in its stable state at the resting point $(u = 0, v = 0)$ until a perturbation displaces the state beyond a precise minimal threshold. If the perturbation is sufficiently large, the state is pushed out of its equilibrium and is entrained by the vector

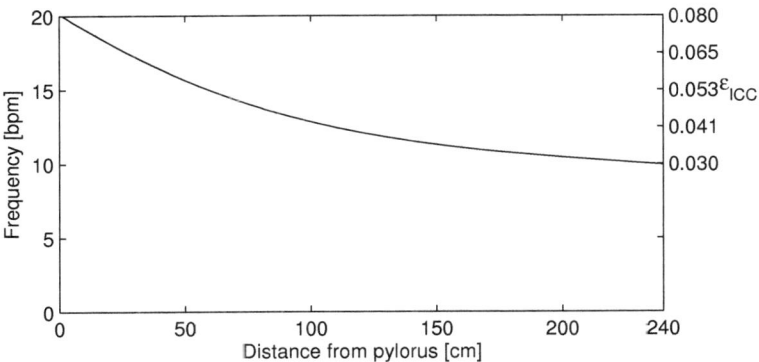

Figure 4.12: Frequency gradient along the small intestine measured and fitted by Aliev. The left axis shows the frequency values in beat per minutes with respect to the distance from the pylorus, while the right axis indicates the corresponding values of ε_{ICC} used in the ICC layer for generating oscillations at these frequencies.

fields along a particular trajectory counterclockwise. The excursion of the state terminates at the resting point $(u = 0, v = 0)$, where the system reacquires its stable state. The figure 4.11(B) displays the corresponding pulse solution. The fields u and v are initially at their resting values, when a perturbation at time $t = 2$ s generates the pulse characterized by an upstroke and a smooth returning to the resting value.

Figure 4.11(C) shows the phase space (u, v) of the ICC layer. It is characterized by a single stable fixed point at $(u = 0.5, v = 0)$, where the nullclines intersect. Due to the fact that the intersection point on the cubic nullcline is located on its middle branch, the dynamics of the system is oscillatory (Hopf–Turing type). This is achieved by adjusting the parameter β of equation 4.60, which is responsible for the shifting of the intersection point from the stable branch of the nullcline (LM layer) to the nonstable one. In fact, the state turns clockwise around the intersection point along an isocline of the vector field defined by the system equations. Should the system not be at the resting point $(u = 0, v = 0)$, in its initial stable state, the oscillatory regime is installed. The corresponding pulse solution is displayed in figure 4.11(D), where the fields u and v are characterized by a periodic wave pattern of frequency 17 bpm.

Intrinsic Frequencies in Isolated Cells

In the ICC layer, the parameter called ε is location–dependent and denoted by $\varepsilon_{ICC}(i)$ (see table 4.3). In a certain range of values, ε is also responsible for the pulsation frequency of the oscillatory system. This dependence allows one to reflect the experimentally observed frequency gradient along the intestine. Fitting was performed on experimental values of the small intestine by Aliev [27] and is depicted in figure 4.12. The gradient has an exponential shape, starting with a frequency of 20 bpm, followed by a rapid decrease that gradually flattens towards the end of the model at 240 cm from the pylorus, where the frequency is 10 bpm. The figure 4.12

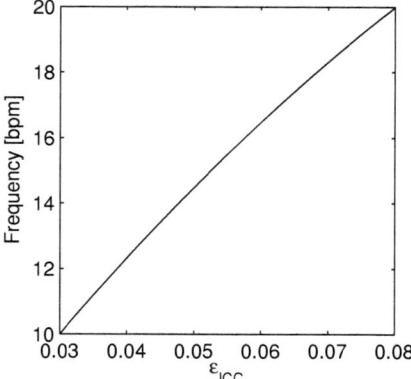

Figure 4.13: Relation between frequency and ε_{ICC} used for the model.

indicates the corresponding values of the ε_{ICC} along the frequency in bpm on the left axis.

The correspondence between the two quantities, i.e. the frequency and ε_{ICC}, is therefore not direct. In fact, the Aliev approach is nondimensional, that is the equations 4.57 and 4.60 of this simple model use two dimensionless state variables (u, v) to represent its dynamics. The parameter values given in table 4.3 are presented without unit, for they are simply mathematical factors chosen to create a particular system behaviour. If a relation with real quantities is requested, a correspondence rule between parameter values and related real values has to be found. In particular, let us focus on the time variable scaling, that is fundamental to fit the model oscillation frequencies with the experimental ones.

The dominant frequency of an isolated ICC cell was evaluated by simulations of different ε_{ICC} and scaled to fit experimental measurements. Furthermore, in his paper, Aliev considers that the value $\varepsilon_{\text{ICC}} = 0.03$ has to correspond to 10 bpm and $\varepsilon_{\text{ICC}} = 0.08$ to 20 bpm. The last step to obtain the time scaling is to make the link between the first point of the fitted frequency evolution in the interval $[0.03, 0.08]$ with 10 bpm and the last point with 20 bpm. Figure 4.13 shows the link existing between the ε_{ICC} oscillation parameter and the desired real frequency in beats per minute. It is interesting to observe that the characteristics of this relation is practically linear, permitting a simplification in the determination of the ε_{ICC} values when creating the desired frequency gradient.

Frequencies in Coupled Cells

So far, only the properties and the comportment of the isolated cell have been treated. Let us consider now the system consisting of a population of bidirectionally coupled LM–ICC pairs of cells, as described at the beginning of this section, and also the role of diffusion process and its related factors.

Each of the two layers is affected differently by the current from the other layer due to the different intrinsic dynamics of the LM and ICC cells. Actually, the last term of the equation 4.57

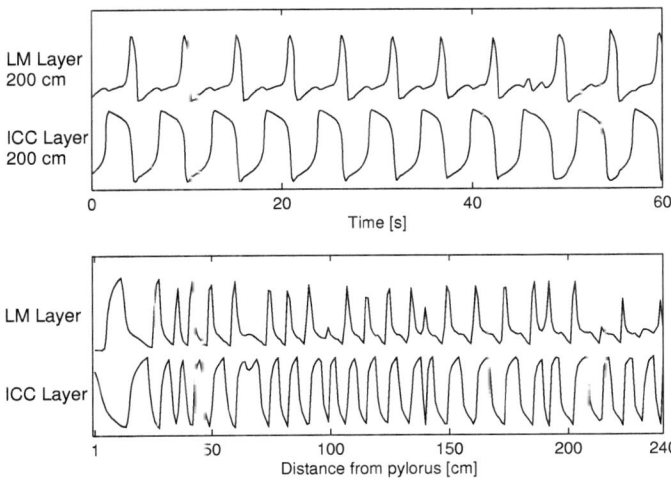

Figure 4.14: Above, profile of the transmembrane potential of the LM–ICC pair of cells located 200 cm from the pylorus. The ε_{ICC} parameter value of the ICC cell is about 0.032. Below, profile of the transmembrane potential along the intestine in the longitudinal muscle and in the interstitial Cajal cells layers.

describes the current density between the two layers. The values of the coupling coefficients D_1 and D_2 are showed in table 4.3. They are inversely proportional to the resistivity of the membrane contacts, so as to mimic a strong coupling inside the LM layer and a weaker coupling inside the ICC layer and between the two layers. Figure 4.14 shows the time and spatial distributions of the transmembrane potential of the LM and the ICC layers for the coupled situation. The evolution of the transmembrane potential at a specific location over 60 s of time, depicted in the upper part of figure 4.14, induces some remarks. The frequency of 11 bpm observed at 200 cm from the pylorus, i.e. the upper figure 4.14, corresponds to the intrinsic frequency of the specific oscillator representing this intestinal segment as shown in figure 4.12. Pulses of ICC layer lead the pulses of the LM layer by several seconds, suggesting that the oscillations initiated in the ICC layer excite the LM layer, where oscillations are generated and propagate. The waveforms are different, in particular the ICC pulses are of different duration, and the LM pulses are inhomogeneously spaced. Note that at time 45–50 s, the dynamics of the LM layer is perturbed, which increases the duration of the ICC pulse and at the same time the inter–pulse in the LM layer.

In the lower part of figure 4.14, the situation of the LM and ICC transmembrane potential along the whole model at a precise instant is displayed. Once again, one may observe that the ICC waves propagating from the proximal part towards the distal part lead of those of the LM layer. Furthermore, not all the ICC waves generate a wave in the LM layer, especially in the distal part. This behaviour can be due to various causes, one of them being a premature excitation of the LM layer. In fact, once excited, the tissue enters a particular state before becoming excitable again. This phase of transmembrane dynamics is well known and studied in cardiac models [20]–[21]. It is named the refractory period.

4.4 Conclusion

In this chapter, we implemented three computer models to simulate the intrinsic electrical activity of the intestine:

Bardakjian: Coupled oscillators. The human colon is represented as a set of 99 coupled oscillators on a tubular structure forming 33 rings of 3 oscillators. Each oscillator represents a small region of the colon and is connected bidirectionally to its 4 neighbors. The values of the intrinsic parameters are chosen in such a way to reproduce observed *in vivo* features of the colon.

Miftakhov: Cellular model based on ionic currents description. This complex mathematical model of the myoelectrical activity of a small intestine functional unit is based on real morphological and electrophysiological data. The kinetics of L–type Ca^{2+}, T–type Ca^{2+}, Ca^{2+}–activated K^+ and Cl^- channels determine the electrical activity of the functional unit. This model gives interesting results in pharmacological experiments with single functional units, but is computationally too heavy for larger scale simulations.

Aliev: Cellular model of the intestine based on FHN representation. This representation is widely used to simulate the dynamics of biological systems. It is constituted of two coupled layers of longitudinal muscle and interstitial Cajal cells. The dynamics of each layer is described using the FHN formalism. The values of the parameters in the model were chosen so that the longitudinal muscle layer is in the excitatory state, while the interstitial Cajal cells layer is in the oscillatory state. Results presented in this chapter were found using a model with 300 cells (as much as Aliev proposed in his paper), namely 150 for each layer.

From the computational point of view, the Bardakjian model has the big advantage to be simple, but this advantage could also become a drawback. If too simple, a model may fail to capture the salient features of the real system and have limited predictive ability.

The realisation of realistic and complex models has become feasible because of recent progress on experimental techniques and computational power. The ionic membrane model of Miftakhov provides the most elaborate and realistic way to model the slow waves and the spike bursts of the intestinal smooth muscle. However, the detailed description of the ionic flows through the cellular membrane implies, unfortunately, a complex formulation, becoming in our case computationally intractable.

The computational cost of adding such a huge amount of details indeed inhibits the use of such models. It is therefore usually appropriate to make some approximations to reduce the model to a sensible size. The engineer can often accept some approximations in order to get a more robust and simple model. This is the main reason for choosing to implement the interesting model proposed by Aliev.

Chapter 5

Numerical Methods

In the field of engineering, one often comes across physical or natural phenomena that have complex mathematical formulations. Finding analytical solutions for such problems becomes difficult, tedious, and is generally even not possible. Here lies the importance of numerical methods, a very powerful tool for getting good estimates of solutions for a wide range of situations.

The physics of the intestine and many other organs is complex. Geometry, structure, and boundary conditions are often irregular, three–dimensional, non–homogeneous, and time varying. Constitutive properties and reaction kinetics are typically non–linear and time dependent. Fundamental physiological functions include mechanical responses and electrical chemical, thermal, and transport processes in cell tissues. Therefore, computational methods are needed to realistically model many of these diverse and heterogeneous processes and their interactions encountered in electrophysiology, biomechanics, and tissue engineering.

For many applications in intestinal physiology, the dynamics of the electrical activity within the cell and the functional interactions can be mimicked by system models typically consisting of coupled sets of nonlinear ordinary differential equations, such as reaction–diffusion equations. Thus, a spatial and temporal discretization is required for the numerical simulation of the time–varying and spatially extended system model describing the intestinal electrical activity. This also necessitates a computational algorithm that is able to track the system evolution at each time step. The way this spatio–temporal discretization is handled leads to different algorithmic solutions that in turn lead to major differences in the final simulation results. The accuracy of the results and the computational complexity, usually usual limited by computational requirements and time needed to achieve a given simulation, are certainly the most important factors. Precision and complexity are two antagonistic properties: the finer the results, and the bigger the complexity. A compromise has to be made, in order to obtain satisfactory simulation results in an acceptable time period.

The discrete handling of time and space always introduces errors, which have to be controlled since they may propagate through time and be strongly amplified by the non–linear dynamics involved in the model. Depending on the temporal and spatial discretization used, some schemes fail to reach any solution at all because of loss of numerical stability as the simulation progresses in time.

This chapter concerns the discretization methods involved to solve numerically the partial ordinary differential equations governing the Aliev model [27] presented in the previous chapter. Temporal and spatial discretizations are considered separately for a better readability. The first section of this chapter presents how time integration can be performed, while the second section presents the spatial discretization schemes. Finally, in the third and last section, the time and spatial discretizations are applied to the specific model of Aliev and the effects of some numerical errors are described.

5.1 Time Integration

Since most ordinary differential equations are not solvable analytically, numerical integration is the only way to obtain information about the trajectory defined by these equations. The model proposed by Aliev [27] and presented in the previous chapter is governed by two ordinary differential equations that can be written in their general form as follows:

$$\frac{du}{dt} = f(u,v,t), \qquad (5.1)$$

$$\frac{dv}{dt} = g(u,v,t), \qquad (5.2)$$

where $f : (u,v,t) \in \mathbb{R} \times \mathbb{R} \times \mathbb{R}^+ \mapsto f(u,v,t) \in \mathbb{R}$ and $g : (u,v,t) \in \mathbb{R} \times \mathbb{R} \times \mathbb{R}^+ \mapsto g(u,v,t) \in \mathbb{R}$ are two differentiable three-variable functions. It is known that sufficient conditions for a unique, continuous, differentiable trajectory (u,v) to exist as a solution to this problem are that $f(u,v,t)$ and $g(u,v,t)$ be defined continuous and satisfy a Cauchy–Lipschitz condition in $\mathbb{R} \times \mathbb{R} \times \mathbb{R}^+$.

Choosing an integration method for extrapolating the state variables is an extremely important task. We need to ensure that the system is stable over time as well as physically correct while keeping the storage requirements and computational costs down. Many methods have been proposed and used universally to solve accurately various types of ordinary differential equations [30], i.e. the Newton, Euler, Crank–Nicholson, Heun, Runge–Kutta, and Adams–Moulton schemes. All these methods discretize the differential equation governing the original system to obtain a difference equation. Of course, the methods produce different solutions from the same differential equation, but they have the same aim: the dynamics of the discretized system should correspond as close as possible to the dynamics of the continuous one.

Numerical discretization of equations 5.1 and 5.2 is performed through a time discretization $t = 0, \Delta t, 2\Delta t, 3\Delta t \ldots$ and an iterative scheme of the form:

$$\mathbf{u}^k \mapsto \mathbf{u}^{k+1}, \qquad (5.3)$$

where $\mathbf{u}^k = (u^k, v^k)$ is an approximation of $\mathbf{u}(k\Delta t) = (u(k\Delta t), v(k\Delta t))$.

Several time integration strategies exist for the evaluation of $\mathbf{u}^{k+1}$, leading to distinct numerical schemes whose characteristics in terms of efficiency, convergence, and stability are sensibly different. The well-known Euler schemes has been widely used for the resolution of ordinary differential equations, since it is simply implemented and provides moderate accuracy for short computation time. [30].

The forward Euler scheme [31] is an explicit method because it uses a combination of past values to evaluate the future state of the system:

$$\frac{\mathbf{u}^{k+1} - \mathbf{u}^k}{\Delta t} = \mathbf{f}\left(\mathbf{u}^k, k\right) \quad \text{and} \quad \mathbf{u}^0 = \mathbf{u}(t_0), \qquad (5.4)$$

where the function $\mathbf{f} = (f,g)$. The stability of this scheme is not guaranteed and, depending on the chosen Δt and the characteristics of $\mathbf{f}$, the algorithm may fail to converge.

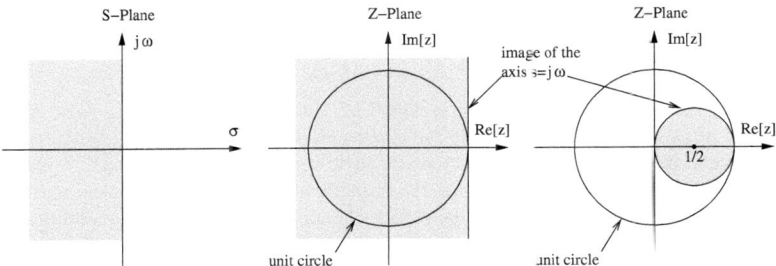

Figure 5.1: Mapping of the s–plane (on the left) onto the z–plane using two different transformations obtained from the forward (in the middle) and the backward (on the right) Euler schemes.

The backward Euler scheme [31] needs a combination of past and future values for the estimation of the future state:

$$\frac{\mathbf{u}^{k+1} - \mathbf{u}^k}{\Delta t} = \mathbf{f}\left(\mathbf{u}^{k+1}, k+1\right) \quad \text{and} \quad \mathbf{u}^0 = \mathbf{u}(t_0). \tag{5.5}$$

The most important advantage of this scheme is that it is unconditionally stable. Since it is an implicit scheme, the evaluation of $\mathbf{u}^{k+1}$ has to be calculated using numerical methods (e.g. Newton).

A linear mixture of the forward and backward Euler schemes [31] is also possible:

$$\frac{\mathbf{u}^{k+1} - \mathbf{u}^k}{\Delta t} = (1-\alpha)\mathbf{f}\left(\mathbf{u}^k, k\right) + \alpha \mathbf{f}\left(\mathbf{u}^{k+1}, k+1\right) \quad \text{and} \quad \mathbf{u}^0 = \mathbf{u}(t_0) \tag{5.6}$$

with $0 < \alpha < 1$, the value $\alpha = 1/2$ corresponds to the Crank–Nicholson scheme. The fully explicit case is defined by $\alpha = 0$, the fully implicit case by $\alpha = 1$, and $0 < \alpha < 1$ provides intermediate semi–implicit cases between these bounds. The scheme is unconditionally stable when $\alpha \geq 1/2$.

When transforming the ordinary differential equations governing a system from the continuous to the discrete domain, it is essential that the principal properties of the system be conserved. The Laplace transform is the analogue equivalent of the Z transform for the discrete domain. If an application between the two planes is established, the conservation of the properties can be studied [32]–[33]. It would be interesting, from the point of view of the stability, that the left semi–plane of the s–plane be applied inside the unitary circle in the z–plane. This essential condition ensures that a stable analogic system is transformed into a stable discrete system.

Consider a linear and time–invariant analogic system characterized by an ordinary differential equation. The first derivative in the continuous domain is represented by a simple difference in the discrete domain. According to the schemes described above, several different applications from the s–plane to z–plane can be obtained. The forward Euler scheme establishes the following correspondence:

$$\frac{d\mathbf{u}(t)}{dt} \longleftrightarrow \frac{\mathbf{u}(k+1) - \mathbf{u}(k)}{\Delta t}. \tag{5.7}$$

The application between the two domains is obtained using the well–known properties of the Laplace and Z transforms:

$$s = \frac{z-1}{\Delta t} \quad , \quad z = 1 + s\Delta t. \tag{5.8}$$

The imaginary axis $s = j\omega$ in the s–plane is shifted by 1 in the z–plane $z = 1 + j\omega\Delta t$, thus the left semi–plane ensuring the stability in the continuous domain is also shifted to the right by 1, as shown in the figure 5.1. It contains the unit circle ensuring the stability in the z–plane, nevertheless such a scheme does not guarantee the stability in the discrete domain since nothing in the iterative process prevents a pole to lie outside the unit circle. Thus, if no precaution is taken in advance, a stable system can be transformed into a non–stable system.

First of all, the time step Δt has to be adjusted conforming to the sampling theory. According to the relation between the Laplace and Z transforms, aliasing effects do not take place if the poles of the system in the s–plane are inside the interval $[-j/\Delta t, j/\Delta t]$. A decrease of the time step value causes an enlarging of this interval that has to be sufficiently extended to contain all the poles.

Another important effect of the value of Δt emerges from the equation 5.8. When the time step becomes smaller, $|z|$ converges to 1. In other words, the system poles in the z–plane are attracted around the point $(1,0)$. The shorter the time step the nearer to $(1,0)$ the poles are mapped. Therefore, when it is fine adjusted the poles are all inside the unit circle and the discretized system becomes stable.

The stability of the system is not the only important aspect to be considered during the discretization process. It is obviously indispensable that the discretized system maintains its stability, but it is also important that its behaviour remains as close as possible to the original. The discretized system has to provide solutions that converge to the values provided by the continuous one. The time step Δt play a fundamental role in this case too, it is responsible of the accuracy of the system. It is well–kown that the Fourier transform is the Laplace transform evaluated along the the imaginary axes in the s–plane and that the discrete Fourier transform is the Z–transform evaluated along the unitary circle. Thus, a decreasing of the time step generates an enlarging of the interval in the s–plane and a compression of the discrete Fourier transform around $z = 1$. An accurate discretization of the transfer function of the analogic system is only possible near $z = 1$, because of the similar tangents assumed by the unit circle and the transformed imaginary axis of the s–plane.

Consider now the map related to the backward Euler scheme that is derived from the following correspondence:

$$\frac{d\mathbf{u}(t)}{dt} \longleftrightarrow \frac{\mathbf{u}(k) - \mathbf{u}(k-1)}{\Delta t}. \tag{5.9}$$

In a similar way as before, the map between s and z planes is obtained through:

$$s = \frac{1 - z^{-1}}{\Delta t} \quad , \quad z = \frac{1}{1 - s\Delta t}. \tag{5.10}$$

The transformation of the s–plane imaginary axis $s = j\omega$ in the z–plane is not directly seen, but a simple expression can be deduced using appropriate substitutions:

$$\left(\text{Re}\,[z] - \frac{1}{2}\right)^2 + \text{Im}^2\,[z] = \frac{1}{4}. \tag{5.11}$$

It is the equation of a circle with a radius of $1/4$ centred in $z = 1/2$, and the left semi–plane in the s–plane is applied inside this circle in the z–plane, as illustrated in the figure 5.1. That application ensures the condition of stability transforming the left semi–plane of the s–plane in the unitary circle of the z–plane. This is the reason why backward Euler scheme is unconditionally stable. Anyway, the value of the time step Δt remains fundamental for the accuracy of the system and the agreement with the sampling theory.

Other applications provide transformations of the left semi–plane of the s–plane exactly in the unitary circle of the z–plane. The Cranck–Nicholson scheme gives rise to a transformation of this type and has been tested by simulations. More generally, the mixed forward and backward Euler scheme is unconditionally stable when $\alpha \geq 1/2$ since the resulting transformation of the left semi–plane of the s–plane are circles inside the unit circle limiting the stability. In fact, for $\alpha = 1$, corresponding to the backward Euler scheme, the left semi–plane of the s–plane is transformed into the circle with a radius of $1/4$ centred in $z = 1/2$. Decreasing the value of α enlarges the disk, since for $\alpha = 1/2$, corresponding to the Cranck–Nicholson scheme, the left semi–plane of the s–plane is transformed into the unit circle. In this case, the stability and the accuracy are of course guaranteed and optimal. A value of α smaller than $1/2$ produces obviously the loosening of those two important properties, as in the limit case of $\alpha = 1$, corresponding to the forward Euler scheme.

5.2 Spatial Discretization

The topic of this section is the discretization of the diffusion equation with no–flux boundary condition:

$$\frac{\partial u}{\partial t} = f(u) + D\nabla^2 u. \tag{5.12}$$

A simple and efficient method for numerically solving partial differential equations is provided by finite differences schemes [30]–[34]. The method consists in evaluating the function only at some specific locations on the domain, replacing the partial differential equations by difference equations.

A finite difference scheme is said to be consistent if it approaches in its limit, as the spatial discretization tends to zero, the original partial differential equation. Consistent schemes can be obtained with the following considerations:

$$\frac{\partial u(x)}{\partial x} = \lim_{\Delta x \to 0} \frac{u(x + \Delta x) - u(x)}{\Delta x} \tag{5.13}$$

$$= \lim_{\Delta x \to 0} \frac{u(x) - u(x - \Delta x)}{\Delta x} \tag{5.14}$$

$$= \lim_{\Delta x \to 0} \frac{u(x + \Delta x/2) - u(x - \Delta x/2)}{\Delta x} \tag{5.15}$$

Consider a cable of length L regularly discretized. The grid is composed of n points and $\Delta x = L/n$. The discrete potential u_i gives an approximation to $u(i \cdot \Delta x, t)$ with $i = 1 \ldots n$ and the diffusion tensor D is considered constant. Forward, backward and centered consistent finite differences are thus defined as:

$$\delta_f u_i = u_{i+1} - u_i, \tag{5.16}$$

$$\delta_b u_i = u_i - u_{i-1}, \tag{5.17}$$

where $\delta_f u_i / \Delta x$ and $\delta_b u_i / \Delta x$ are the forward and backward approximations of the first order partial derivative respectively. The difference operators help build approximations of higher order differential operators, and the mth order finite difference can be obtained by recursively applying one of the three operators:

$$\delta_f^m u_i = \delta_f\left(\delta_f^{m-1} u_i\right), \tag{5.18}$$

$$\delta_b^m u_i = \delta_b\left(\delta_b^{m-1} u_i\right), \tag{5.19}$$

where $\delta^m u_i/\Delta x^m$ can be considered as an approximation of the mth order partial derivative.

The schemes for the second order difference operator are obtained by successively applying the first order difference operator. The centered second order finite difference can also be obtained as:

$$\delta_c^2 u_i = \delta_f \delta_b u_i = u_{i+1} - 2u_i + u_{i-1}. \tag{5.20}$$

The 1D approximation diffusion operator becomes:

$$\frac{1}{\Delta x^2}\delta_c^2 u_i = \frac{u_{i+1} - 2u_i + u_{i-1}}{\Delta x^2}. \tag{5.21}$$

After an adequate spatial discretization using a constant space step Δx, the evolution of the transmembrane potential at any point i can be explicitly updated from time k to time $k+1$, and equation system 4.57–4.60 becomes:

$$u_{\text{LM},i}^{k+1} = u_{\text{LM},i}^k + \Delta t \left(\kappa u_{\text{LM},i}^k \left(u_{\text{LM},i}^k - a\right)\left(1 - u_{\text{LM},i}^k\right) - v_{\text{LM},i}^k\right) +$$
$$+ \frac{\Delta t D_1}{\Delta x^2}\left(u_{\text{LM},i-1}^k - 2u_{\text{LM},i}^k + u_{\text{LM},i+1}^k\right) +$$
$$+ \Delta t D_2 \left(u_{\text{LM},i}^k - u_{\text{ICC},i}^k\right), \tag{5.22}$$

$$v_{\text{LM},i}^{k+1} = v_{\text{LM},i}^k + \Delta t \left(\varepsilon\left(\gamma\left(u_{\text{LM},i}^k - \beta\right) - v_{\text{LM},i}^k\right)\right), \tag{5.23}$$

$$u_{\text{ICC},i}^{k+1} = u_{\text{ICC},i}^k + \Delta t \left(\kappa u_{\text{ICC},i}^k \left(u_{\text{ICC},i}^k - a\right)\left(1 - u_{\text{ICC},i}^k\right) - v_{\text{ICC},i}^k\right) +$$
$$+ \frac{\Delta t D_1}{\Delta x^2}\left(u_{\text{ICC},i-1}^k - 2u_{\text{ICC},i}^k + u_{\text{ICC},i+1}^k\right) +$$
$$+ \Delta t D_2 \left(u_{\text{ICC},i}^k - u_{\text{LM},i}^k\right), \tag{5.24}$$

$$v_{\text{ICC},i}^{k+1} = v_{\text{ICC},i}^k + \Delta t \left(\varepsilon\left(\gamma\left(u_{\text{ICC},i}^k - \beta\right) - v_{\text{ICC},i}^k\right)\right) \tag{5.25}$$

where the equation parameters are defined in the previous chapter.

5.3 Numerical Instabilities and Errors

Numerical simulations of physical or natural phenomena obtained by discretization of partial differential equations expressing variations of physical values in space and/or time always introduce numerical instabilities and errors [31]. In the particular case of the electrical propagation in the intestine it is important to have a close control on these effects since they may propagate through time and be strongly amplified by the highly non linear dynamics involved in the membrane reaction. Considering that the time discretization is appropriately selected for the implemented scheme, these two negative effects are principally related to spatial discretization. Numerical instabilities require large space steps to be avoided while numerical errors become acceptable if a very fine grid is used. In order to find an adequate grid scale, the two antagonistic effects have to be minimized jointly.

Stability Criterion

Numerical instability takes place when the diffusive part of the propagation equation is computed explicitly from one time step to the next. It is related to the propagation of information to a finite neighbourhood around each spatial node from one time iteration to another.

The stability of explicit diffusion algorithms computed on a regular grid is a well known problem, initially investigated in heat conduction and chemical diffusion processes. As a basis of the discussion, recall the diffusion equation in one space dimension of the parabolic model when the diffusion coefficient D is considered constant:

$$\frac{\partial u}{\partial t} = D \frac{\partial^2 u}{\partial x^2}. \qquad (5.26)$$

This can be discretized in the following way:

$$\frac{u_i^{k+1} - u_i^k}{\Delta t} = D \frac{u_{i+1}^k - 2u_i^k + u_{i-1}^k}{\Delta x^2} \qquad (5.27)$$

Using the Von–Neumann analysis [31], the criterion for homogeneous and isotropic diffusion process computed on a regularly discretized and infinite grid using finite differences is:

$$\frac{D \Delta t}{\Delta x^2} \leq \frac{1}{2d}, \qquad (5.28)$$

where d is the dimension of the tissue ($d = 1$ for a 1D cable, $d = 2$ for a 2D patch, and $d = 3$ for a 3D block of tissue). The physical interpretation of the restriction expressed by equation 5.28, coupling temporal and spatial discretization, is that the maximum time step allowed is, up to a numerical factor, the diffusion time across a cell of width Δx.

As Δt should be taken as the maximum value allowing a correct integration of the discretized equations to ensure a minimal number of iterations of the algorithm, a typical minimal time step guaranteeing the numerical stability of the iterative scheme (equations 5.22 and 5.25) can be determined using equation 5.28. With $\Delta t = 10$ ms, and the standard values for the worst case $D = 0.4$, the minimal value for Δx in a 1D tissue is:

$$\Delta x \geq 1.6 \text{mm} \qquad (5.29)$$

The stability criterion shows that stability problems are to be expected in tissues having large diffusion factors, in multidimensional tissues, and mostly in finely discretized grids. If any constraint imposes a space step smaller than allowed by equation 5.28, as the minimization of the numerical errors associated to spatial discretization might do, the only way to stay within the stability domain is to decrease Δt. Because of the quadratic relation between Δx and Δt, such a scenario should be strongly avoided because of its huge impact on the number of time iterations required to complete a simulation of fixed duration.

The criterion expressed by equations 5.28 fixes the inferior limit of the space step, but it is not the only restriction to be respected. In fact, a superior limit arising from the inspection of numerical errors has also to be considered. A qualitative way to choose the appropriate space step and a means to estimate the unavoidable errors introduced by the discretization process is to observe the variations of the conduction velocity in the tissue. Conduction velocity has been chosen because its alteration is a typical and easily recognizable signature of coarse discretization.

Two waves propagate in two tissues differently discretized with a different velocity, and when Δx tends to zero, the conduction velocity approaches its limit value, which is ordinarily equal

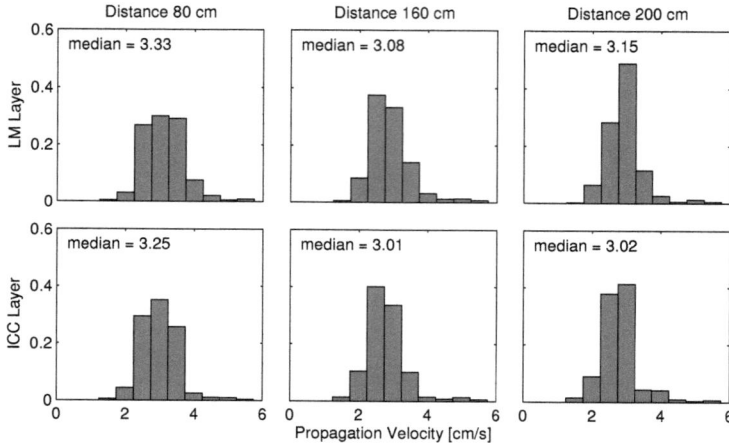

Figure 5.2: Normalized distribution of the wave propagation velocity in the LM and ICC layers at different positions along the Aliev model. The space discretization is $\Delta x = 8$ mm, while the time step is $\Delta t = 10$ ms.

to the real value. The discretization process is usually considered acceptable if the relative error affecting the conduction velocity is lower than 10% [31]. The approach used here is quite different from the typical ones described in the literature, because of the nature of the Aliev model. In excitable tissues, such as the heart models, propagation waves can be generated with a current stimulation and conduction velocity is easily evaluated. In the case of the intestinal model, many waves propagate simultaneously in both directions along the cable tissue and are self-generated at random positions and instants.

In order to evaluate the conduction velocity at a fixed position along the cable, a single wave has to be detected and tracked over an established distance. This operation is performed during a certain period of time where various waves are tracked and the distribution of their propagation velocities is evaluated and displayed using an histogram. Figure 5.2 shows the distribution of the propagation velocities evaluated at three different positions along the two layers of the cable. The six histograms indicate similar distributions, where propagation velocities are grouped around a similar value. This is the sign that the propagation velocity is stable in time because of its relatively small range of variations. A representative value of the wave propagation velocity at a precise location can be identified using one of the typical quantities widely used in statistical analysis. The median value, also displayed above each histogram in figure 5.2, is chosen as a representative propagation velocity. The three considered regions of both layers exhibit a similar distinctive median value of about 3 cm/s, and then probably, the propagation velocity median is the same for all the regions of the cable. In order to clear the suspicion, the tissue characteristics can be monitored all along the cable.

Median values of the wave propagation velocities are calculated along the whole model and displayed with respect to the positions in figure 5.3. Except for the first section of the cable, from

Section 5.3   Numerical Instabilities and Errors   61

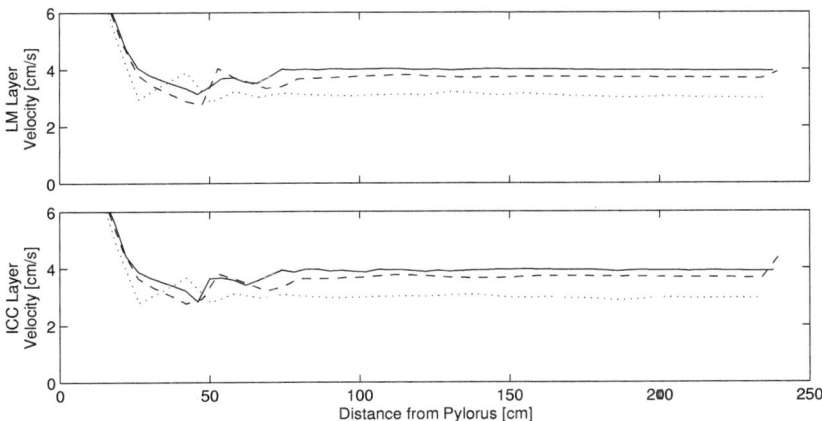

Figure 5.3: Medians of the wave propagation velocity along the Aliev model for three different spatial discretization values $\Delta x = 8, 5, 4$ mm, which correspond respectively to the dotted, dashed and solid lines.

position around $x = 50$ cm to the end the median of the propagation velocity is constant. At the beginning of the cable, the strong synchronism between the fastest oscillators together with some boundary effect imposes a high propagation velocity. Even though the values are realistic after the stability is reached at about 50 cm, the wave speed in the first part of the tissue is too high. This unpleasant behaviour has to be taken into account during further measurements so as to know the lack of realism of the model and then control its limits. However, the order of magnitude of the propagation velocity along the rest of the tissue is coherent with physiological observations.

Nevertheless, as also shown in figure 5.3, the velocity is strictly dependent on the spatial discretization used. Three different space steps are used, $\Delta x = 8, 5, 4$ mm, so as to give an idea of the incidence they have on the observed propagation velocity. These three values are of course consistent with the stability criterion of equation 5.28, but the numerical errors associated to each spatial discretization influences differently the results. In general, the smaller the space step the smaller the numerical errors. Thus, the fact that in figure 5.3 the decrease of the space step gives rise to a monotonous increase of the conduction velocity, is a comforting sign of convergence of the numerical error. Besides, it is also important to note that both layers exhibit the same behaviour. All the previous observations are then valid for both.

It is useful to inspect the evolution of the numerical errors in depth, in order to identify an interesting and acceptable value for the space step. Supposing that the median of the wave propagation velocity is stable and constant at all positions for a specific space step, the observed quantity can be monitored for different spatial discretizations at a certain position, considered as representative of all the cable   Figure 5.4 shows the evolution of the median conduction velocity with respect to the number of oscillators in one layer. The medians are evaluated at the representative position $x = 150$ cm. Both layers exhibit a similar feature, that is, a considerable

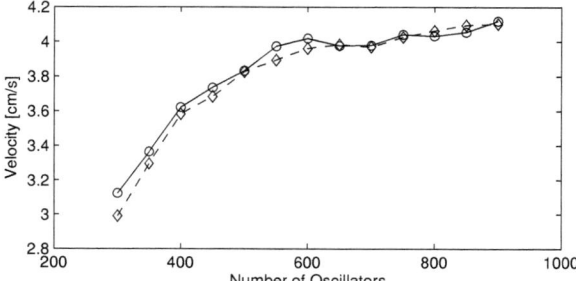

Figure 5.4: Median of the wave propagation velocities at position $x = 150$ cm with respect to the number of oscillators composing one layer of the Aliev model. Solid and dashed line correspond respectively to LM and ICC layers.

increase of the velocity for the model composed of less than 600 oscillators for each layer, and a stabilization of the velocity at about 4 cm/s for a finer discretization. The increase of the number of oscillators in the model corresponds of course to a decrease of the spatial discretization. This particular representation has been preferred since the convergence effect is clearly visible and then a suitable spatial discretization value is easily identifiable.

As shown before, the minimal space step guaranteeing the numerical stability of the iterative scheme is $\Delta x \geq 1.6$ mm, but at the same time, a coarse spatial discretization is not acceptable because of the numerical errors affecting the conduction velocity. The space step $\Delta x = 4$ mm corresponding in figure 5.4 to 600 oscillators, gives satisfactory results from the point of view of numerical errors. In fact, using this spatial discretization, the relative error affecting the conduction velocity is lower than the limit fixed to 10%.

Actually, the control of the conduction velocity of slow waves in the small intestine seems to be more complex. Velocities of slow waves have been measured both in the longitudinal and circumferential directions, and this highlighted an anisotropic propagation in the tissue [35], faster in the circumferential than in the axial direction. Moreover, note that in the Aliev intestinal model, slow waves propagate with a conduction velocity 2–2.5 times greater than the experimental one.

Even if the choice of an adequate time step is of course important as far as keeping numerical errors in check is concerned, this issue will not be discussed in detail here. The errors related to time step discretization are smaller and much easier to control than the ones associated to spatial discretization, thus justifying the emphasis put here. The computational requirements are also very sensitive to the chosen time step since more iterations are required to simulate a given duration with a finer temporal discretization. Contrarily, large time steps influence the up– and down–stroke computation and decrease the conduction velocity of the waves. The time step has also to be chosen with regards to the time derivative of the model variables. In order to show this effect, a time step ten times finer than the standard $\Delta t = 10$ ms has been used, and the relative error between the two conduction velocities was only 1.4%.

5.4 Conclusion

Simulation of physical or natural phenomena usually implies a discrete integration of partial differential equations expressing variations of physical values in time and/or space. In this chapter the basic discretization schemes involved to solve numerically the reaction–diffusion equations are presented.

Forward, backward, and mixed Euler schemes for the time integration are presented and their stability and convergence are discussed. The implicit methods have the advantage of being immune to numerical instabilities, but their high computational cost is still restraining their use. The explicit one (forward Euler) is the simplest and fastest method, while the risk of numerical instability has to be carefully investigated.

A simple and efficient method, based on the centered second order finite difference, has been used to discretize the one–dimensional diffusion equation.

As a result, numerical methods do not usually give the exact answer to a given problem. The discrete handling of time and space always introduces numerical instabilities and errors that can propagate through time and be strongly amplified by non–linear system dynamics. Approximating a continuous tissue by a finite number of spatial nodes introduces errors, typically a reduction of the conduction velocities of slow waves in the tissue. However, it has been shown that temporal and spatial discretization steps are related by a stability criterion that fixes the minimal space step for a given time step. It has also been shown that the reduction of conduction velocity due to spatial discretization is higher than the one introduced by temporal discretization.

To numerically solve equations of the Aliev model, we used the classic explicit Euler method for time discretization, while space was discretized with a one–dimensional discrete Laplacian.

Chapter 6

Relating Simulations to Experimental Signals

Computer models can provide information at multiple biological scales and give access to every region of interest with a high resolution. The previous chapters presented the ways to generate this information and then the precautions to take for obtaining realistic results.

In contrast, clinical evaluation of the human gastro–intestinal electrophysiology usually involves only time–limited and noisy electrical signals measured with unipolar electrodes, serosal or intraluminal electrodes, or an electrode array. In order to get closer to the electrophysiologist's viewpoint, data from simulated intestinal electrical activity will be represented in a form similar to those of experimental and clinical studies. Similar reasons give incentive to develop new approaches to compare simulated data with the characteristics of the movement of intraluminal contents, which are observed with novel techniques used in gastroenterology. This operation definitely reduces the amount of information available but is a necessary step toward model validation.

This chapter presents the description of an ensemble of tools for the intestinal activity in terms of spatial organization, electrograms, and mass movement. For the sake of completeness, the last section is dedicated to the description of three dimensional visualization of the colon. Comparison with published experimental or clinical data will be performed in the next part using the techniques presented in this chapter.

6.1 Space–Time Maps

Space–time maps or plots constitute a useful representation of the behaviour of oscillatory systems. They are built by stacking the states of the system at successive time instants, or a measure of them. Thus, the x–axis of the map represents the spatial position along a system, and the y–axis represents the time. This representation provides important information on the spatial and temporal organization of the system dynamics, and especially the presence of propagation waves along the system, their direction and velocity, the initiation sites of propagation, the collision of forward and backward waves, etc.

This visualization technique is also used to analyse experimental measurements, in which electrical or mechanical activity of the intestine is recorded. In order to analyse the components of the motor pattern in the guinea–pig small intestine (55 mm long), Henning developed a simple method for constructing spatio–temporal maps of changes in diameter and longitudinal muscle length from video recordings [36]. Each single frame of a digital video recording of the intestine is binarized, so as to obtain the profile of the intestine. The diameter of the intestine at regular locations along its length is obtained by counting the number of pixels composing its profile. Multiple frames are treated in this way and the results of consecutive images are stacked on top of each other to produce a spatio–temporal map. Using this technique the profile of the intestine at each position and time can be reconstructed.

Spatio–temporal maps of contractile events in the intestine have also been experimentally measured by Berčík with video images analysis of isolated segments of rat ileum (5–7 cm long), combined with manometry and electromyography [37]. He developed an arterially and luminally perfused rat ileal loop preparation, which shows characteristic myogenic and neurogenic rhythms, and intestinal pacemaker activity. The intestine, when distended and filled with a limpid medium, is semitransparent. Under appropriate illumination, contractions and relaxations can be detected as zones of increased or decreased opacity of the intestinal wall, respectively. Video images are digitized and the pixels along the longitudinal axis of the intestinal segment are extracted from each image and displayed as a straight line. These axial lines from consecutive images are stacked on top of each other to create a spatio–temporal map in which any propagated contraction appears as an oblique light line.

Space–time maps are a useful representation of *in vitro* experiments, although the results are generally perturbed by some element. The greatest limitation of experimental space–time plots is due to the noise injected in the recorded video sequence by several factors: illumination, exposition, focus, digitalization of images, extrapolation of the information from images, and the spatial resolution of spatio–temporal maps determined by pixel density. Moreover, *in vitro* experiments permit to observe only relatively small segments of the intestinal tract.

Results obtained with the Aliev computer model of the intestine can also be displayed using this powerful representation. The transmembrane potential along a small portion of the Aliev model is displayed for successive time instants on the left of figure 6.1. The left of the images always represent the position in the oral direction, and the right the aboral one. Thus, pulses propagating from the left of the image to the right are forward waves, and in the opposite case backward waves. Two of them, corresponding to the backward and forward cases, are highlighted with arrows indicating the direction of the propagation. Pacemaker sites are particular regions where waves are initiated rhythmically during a certain period of time. This phenomenon is also indicated on the left of figure 6.1 where a gray disk highlights the position and the precise instant of initiation of two waves, one propagating orally (backward direction) and the other aborally (forward direction). The second gray disk shows the position and the later instant where the backward wave collides with a wave coming from the opposite direction. After a collision, normally, waves stop propagating because the tissue is in the refractory period in both directions.

The space–time plot of the transmembrane potential for the whole model is obtained similarly. The only difference lies in the coding of the transmembrane potential values with a gray level map, where the black and white corresponds respectively to the maximum and the minimum values. On the right of figure 6.1 the resulting space–time map is depicted, where the x–axis still represents the location along the model and the y–axis the time. Using this visualization, propagation appears as an oblique black line with a slope depending on propagation velocity. The initiation of waves, as well as their direction and velocity of propagation can be easily detected

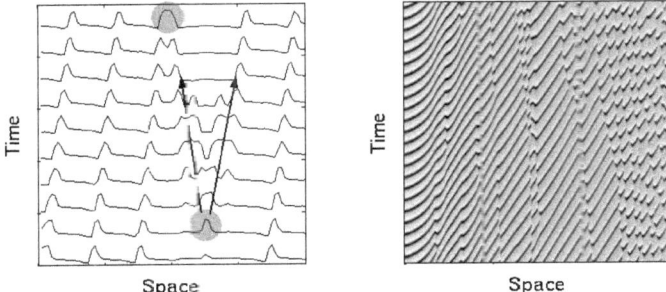

Figure 6.1: Examples of the transmembrane potential in the LM layer, obtained using the Aliev model. On the left, the transmembrane potential along a portion of the model is plotted for 10 successive time instants. Backward and forward propagating waves are displayed (left and right arrows respectively), as well as a pacemaker site (inferior gray disk) and a collision of two waves (superior gray disk). On the right, an example of space–time plot of the whole model. Black and white corresponds respectively to the maximum and the minimum values. Any propagating wave appears as an oblique black line.

and observed.

This type of visualization permits to display relatively extended periods of spatially and temporally complex dynamics in a single plot. However, other important informations on the system, such as intestinal space organization, would be interesting to observe. Spatial organization evolves slowly and needs to be observed for a long time to witness meaningful changes. Thus, a different representation allowing us to observe the evolution of system organization, is needed.

6.2 Synchronization and Space Organization

One of the central challenges facing many scientific disciplines, including physiology, biology, and physics [38], is to understand the organization of complex networks of interconnected components. The studies focusing on the identification of the generic behaviour or the organization principles of such systems often deal with synchrony, a phenomenon in which the dynamics of multiple components becomes simultaneous. The concept of synchronization goes back to the observation of interactions between two pendulum clocks by Huygens. Synchronization of oscillatory systems has been widely studied but it was not until recently that synchronization between chaotic motions drew attention.

Various definitions of synchronization have evolved. In particular, the concept of phase synchrony [39] has proven to be very useful in the analysis of biological data [40]–[41]. In biological systems, synchronization is important since it is the source for an organized activity, which may be essential to their function. It would be interesting to implement a simple and

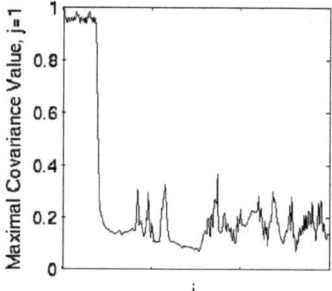 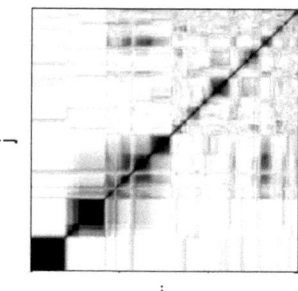

Figure 6.2: Left side represents the maximum of the normalized cross–covariance values of the LM transmembrane potential between cell $j=1$ and all the others. On the right an example of cross–covariance plot is shown. In the matrix M, black corresponds to one, and white to zero. Square–like structures along the main diagonal represent clusters of synchronized cells.

robust method [42]–[43] leading to a direct interpretation of the large amount of data generated with the Aliev computer model by displaying characteristics of the space organization of the intestine and its evolution with time.

6.2.1 Cross–Covariance Analysis

In a stationary situation, or in a time window during which stationarity can be assumed, it is possible to estimate the normalized cross–covariance of the transmembrane potential signals for each pair of cells. Then, one extracts the maximum cross–covariance values, and groups them in a matrix M of dimensions N by N, where N is the number of cells in the LM layer [44]. More precisely, if the dynamics of a single cell is $x_i(t)$, then the lattice at time t is represented as $\{x_1(t), x_2(t), \ldots, x_N(t)\}$. The elements of the matrix M are defined by:

$$M_{i,j} = \max\left(\frac{\phi_{i,j}(k)}{\sqrt{\phi_{i,i}(0)\,\phi_{j,j}(0)}}\right), \tag{6.1}$$

where $\phi_{i,j}(k)$ is the estimated cross–covariance of signals x_i and x_j. The values contained in the matrix M are in the interval $[0,1]$. Moreover, assuming stationarity, the cross–covariance property $\phi_{i,j}(k) = \phi_{j,i}(-k)$ imposes on M a symmetry around the main diagonal $i = j$, where the maximal values are reached.

Clusters of synchronized cells can then be easily identified in a gray level representation of M. Phase synchronous clusters are zones where cell activities tend to be related in some way. They appear like square regions along the main diagonal of M. The larger the square–like structure on the main diagonal of M, the larger the cluster.

Figure 6.2 displays on its left an example of $\phi_{i,j=1}$, obtained using the signals in a time window of a typical Aliev model simulation. The profile $\phi_{i,j=1}$ corresponds to the first line of the matrix M. When all the values are uploaded in the matrix M, it can be displayed as shown on the right of figure 6.2, where black square regions placed on the diagonal represents clusters of synchronized cells. Maximum values are reached along the main diagonal, but remain close to one in the neighborhood of synchronized cells.

In a non-stationary situation, one can estimate the matrix M in a time window of appropriate duration, during which stationarity can be assumed, and then move this window so as to obtain a time–sequence of matrices M_t. We used a window length of 256 samples and a window translation of 64 samples. It seems to be a good compromise between accuracy and time resolution. The resulting 3D matrix sequence is difficult to interpret. Information has to be extracted from each matrix, in order to remove one dimension in the resulting representation.

Let us point out that the method we presented above can be extended without any change to other popular linear and nonlinear measures [45], such as the coherence function, the mutual information, or the Lyapunov exponents. The recurrence plot is an interesting nonlinear technique that permits the analysis of multivariate data [46]. Comparing the dynamics of two time series by testing the closeness in the phase space of each value of two trajectories, the cross recurrence plot can be constructed and shows signal interrelations similarly to the presented cross–covariance analysis. This nonlinear technique is widely used [47]–[48] in applications to heart rate variability data.

In our model the time evolution at each cell remains sufficiently simple and smooth that measures based on second–order statistics are powerful enough to quantify the relationship between the time evolutions of two cells.

6.2.2 Cluster Evolution Analysis

Borders between clusters of synchronized oscillators correspond to the junctions between square–like structures along the main diagonal of M. Thus, an appropriate estimate of the spatial extension of the clusters can be obtained as follows. First, the elements of each matrix M_t are thresholded to zero or one with a value of 0.7 in this specific case. Second, the number of non–zero values along each anti–diagonal is recorded which yields a profile $p_t(k)$, $k = 1 \ldots N$, of the matrix M_t.

Local minima of this profile delimit cluster borders, local maxima define cluster centres, and peak widths are related to cluster sizes. The only remaining step consists in stacking the profiles on top of each other to create a representation of the time evolution of the clusters.

Figure 6.3 shows the three major steps to estimate the cluster sizes using the method presented above. The assessment of the different cluster sizes in the computer simulations can be easily estimated observing the number of values greater than a certain threshold along the anti–diagonals of M, as depicted on the left of figure 6.3. This yields a particular profile illustrated in the middle of figure 6.3. The values of the cluster size estimates at three particular positions of the M anti–diagonal are indicated with three circles.

The profile can be represented with respect to time as shown on the right of figure 6.3 for several minutes of simulated electrical activity. Observation of this visual representation inspires some remarks. First of all, there is a stable cluster at the beginning of the array. A zone of periodic organization follows it. The central part shows a typical pattern of irregular cluster sizes, which become larger at the end of the lattice.

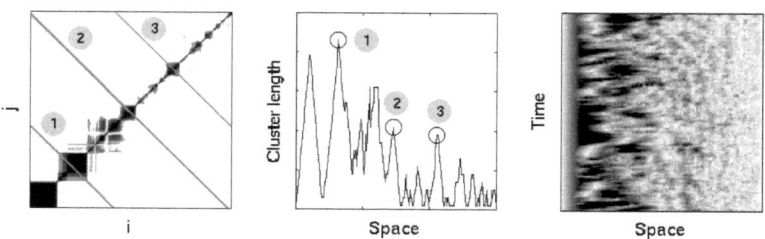

Figure 6.3: On the left, thresholded version of the gray–level representation of a matrix M. In the middle, the corresponding cluster lengths evaluated along the antidiagonal of the matrix M. On the right, evolution of the cluster size, where black corresponds to a value larger than 60 cells and white to zero.

6.3 Electrograms

The measurements of bioelectric potentials has challenged the state of technology since the eighteenth century, and this challenge has led to many improvements in instrumentation and, arguably, in digital computers [49].

Biological tissues are made of cells immersed in a fluid matrix. The cells are surrounded by a membrane through which ions and molecules may pass in either direction. Control of mass transport by the membrane is a major aspect of life processes at the cellular level. In general, the interior of the cell has a negative potential with respect to the exterior. Hence, cells in the resting state are electrically polarized. The change in potential is accompanied by movements of ions across the membrane. Thus, the electrical activity of cells results in currents and electric fields everywhere in the body. Potential differences on the skin of readily measurable amplitude are often created.

Electrical activity of neurons in the brain gives rise to the electroencephalogram (EEG) on the scalp. Activity of skeletal muscle gives rise to the electromyogram (EMG) which may be detected on the skin overlying the muscle. Activity of the heart cells gives rise to the electrocardiogram (ECG). Activity of the gastric smooth–muscle gives rise to the electrogastrogram (EGG).

The ECG is the major diagnostic instrument of cardiac electrophysiology. It is estimated that approximatively 100 million ECGs are recorded annually in the United States. Moreover, the ECG is also displayed during monitoring of patients in the operating room.

For an EGG, several surface electrodes are taped onto the abdomen over the stomach in the same manner as electrodes on the chest for an electrocardiogram. The electrodes sense the electrical signals coming from the stomach muscles, and the signals are recorded on a computer for analysis. The EGG can be considered an experimental procedure since its exact role in the diagnosis of diseases of the stomach has not been elucidated yet. It is used when there is a problem with recurrent nausea and vomiting, signs that the stomach is not emptying food normally. Recordings are made both in fasting and after a meal with the patient lying quietly, and last for two or three hours. If the EGG is abnormal, it confirms that the problem probably is with the stomach muscles or the nerves that control the muscles. In normal individuals there is a regular electrical rhythm generated by the muscles of the stomach, and with an increase in amplitude after the meal. In patients with abnormalities of stomach muscles or nerves, the rhythm is often irregular or there is no post-meal increase in electrical activity. Surface measurements of electrical activity of the small intestine can also be recorded [50].

Cardiologists have found crucial in certain situations to study electrograms directly from the heart by means of a catheter inserted through a vein or artery, or at the time of surgery. In a similar way, invasive methods for recording electrograms of the entire gastro-intestinal smooth-muscle are used. The established method is the use of serosal or intraluminal electrodes [51]. Serosal electrodes are implanted on the serosal surface of the intestinal tract during an abdominal surgical procedure [52], while intraluminal electrodes are placed under endoscopy. The applicability of the serosal method is very limited and most studies can be performed only in animals. The intraluminal method, on the other hand, disturbs the on-going activity of the small intestine.

An invasive technique consisting in multiple simultaneous recordings from a large number of extracellular electrodes is widely used in the study of electrical conduction of the heart [53]. A similar idea is used to investigate the gastro-intestinal electrophysiology of a small patch of tissue [54]. A multiple-electrode assembly consisting of 240 silver wires is positioned on the serosal surface of the intestine, and the electric potentials are recorded. Analysis of the electrical mapping shows the propagation of slow waves [55] and spikes [56] in the two dimensions on the tissue patch.

In computer models, a direct access to membrane potentials is available. Therefore, a comparison with electrical mappings is relatively easy. A unipolar electrode is modeled by a point measurement of the extracellular electric potential, where the reference electrode, thus the zero potential, is located at an infinite distance. In contrast to experimental sensors, this idealized electrode does not damage the tissue or cause disruptions, and does in no way alter the current sources or the propagation of depolarization waves. Moreover, its application on the tissue does not modify the tissue surface geometry.

The extracellular potential can be estimated assuming a superposition of potential fields from each one of the transmembrane current sources [34]. Let I_m be the current per unit surface and ϕ the electric scalar potential at a point in the volume conductor then:

$$\phi = -\frac{1}{4\pi\sigma_{\text{ext}}} \int_\Omega \frac{I_m}{r} \delta ds, \qquad (6.2)$$

where Ω represents the surface of the tissue, ds is an infinitesimal surface element, σ_{ext} is the extracellular conductivity, δ is the thickness of the tissue, and r is the distance from the sources to the observation point. The numerical quadrature evaluation of this integral is performed by lumping the sources. On a grid $\mathcal{G}$ containing points i located at x_i and associated with the region of constant area Ω_i, the approximation reads:

$$\phi(x,t) = \frac{\delta\Omega_i}{4\pi\sigma_{\text{ext}}} \sum_{i\in\mathcal{G}} \frac{I_m(x_i,t)}{\|x-x_i\|}, \qquad (6.3)$$

where the factor $\delta\Omega_i/4\pi\sigma_{\text{ext}}$ is a constant. Since I_m is time-dependent in the right hand side, a unipolar electrogram is computed as a linear combination of current sources. The membrane current I_m is obtained by applying the diffusion operator to the membrane potential field, so that the electrogram be also a linear combination of membrane potentials. Only one parameter characterizes a simulated electrode, namely its distance d to the surface of the intestine. This

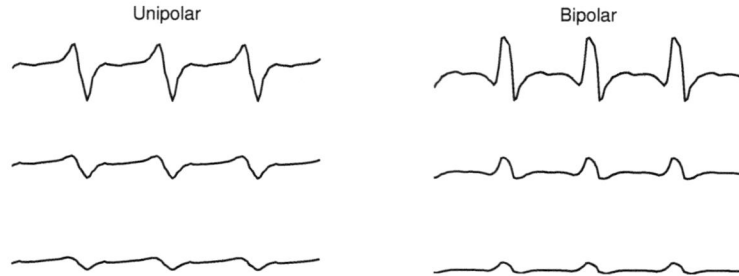

Figure 6.4: Examples of simulated unipolar electrograms, on the left, and 1 cm–spaced bipolar electrograms on the right. Arbitrary units are used for the illustration. The three electrograms are calculated using three different distances of the electrode from the intestine surface, increasing from top to bottom.

parameter can be understood as a parameter controlling the characteristic length of the region the electrode considers for averaging. A bipolar electrode is modeled as the difference in electric potential between two points close to each other, typically a few millimeters, and both at the same distance from the tissue surface.

Figure 6.4 shows unipolar electrograms on the left side and bipolar electrograms on the right side, computed during a typical Aliev computer simulation of the electrical activity of the intestine, taking into account the appropriate changes of the equation 6.3 to be used for a cable model. Three different distances d of the electrode to the intestinal surface are used to obtain the three examples of electrograms in the figure, in order to show the effect of the distance on the electrograms. First of all, in both cases, unipolar and bipolar, the greater the distance the smaller the electrogram amplitude, but also, the electrogram shapes are different. As mentioned above, the distance from the surface is responsible for the length scale of the averaging effect, i.e. the region an electrode can see. The values of d used to obtain the three electrograms are not important, because only the qualitative results are considered. Nevertheless, the distance used to simulate electrodes in contact with the tissue are generally greater than in a real situation, so as to compensate for the fact that the model has an ideal point–electrode. We can consider that the region an electrode can see can correspond to the order of magnitude of the real surface sensors.

An interesting model of skin–surface electrode simulation has been introduced by Familoni in [57] and further developed [58]. He computed the electromagnetic effect of the electrical intestine activity through the human torso, which was assimilated to a finite cylinder of homogeneous medium. Similar models have been used, for example, to evaluate the negative effect of the abdominal thickness of the electrogastrogram using synthesized gastric signals [59], or to compare the electric potentials and magnetic fields that result from intestinal electrical activity [60]. Today, more complex human torso models exist, which consider different discrete regions representing heart, lung, bone, blood, muscle, etc., with an appropriate conductivity assigned to each region.

6.4 Mass Displacement

Many studies, in different ways, have shown the temporal relationship between slow waves, spikes, and muscle contraction [61]. In the small intestine, the slow waves propagate in the organ on a network of interstitial Cajal cells, and smooth muscle cells react by patterns of contraction. Smooth muscle spikes, initiated in response to slow waves, appear to propagate through the smooth muscle layer and are often the first step of the excitation coupling mechanism leading to contraction. Isolated smooth muscle cells do not generate spikes spontaneously; the latter occur only when the membrane potential is depolarized above a specific threshold. Whether or not spikes are initiated seems to depend on the amplitude of the depolarization generated in the smooth muscle cells by the slow waves.

Motility is one of the main functions of the digestive system. The small intestine spontaneously produces two principal motility patterns: mixing movements (segmentation and pendular movements) and propulsion movements (peristalsis). The relationship between these contraction patterns and the different electrical signals are associated is not clear.

The various recognized exploratory and clinical approaches to measure intestinal motility provide interesting results. Indeed, the continuity of luminal content displacement along the gastro–intestinal tract is difficult to study. Transit tests with radio–labeled pellets is a well–established and widely used technique, which provides general information concerning the luminal transport. Intraluminal manometry (perfusion catheters and pressure gauges) contributes in characterizing the rhythmical phenomena and pressure gradients, but does not allow a description of luminal flow patterns and is mainly limited to proximal and distal segments. Imaging methods (X–rays, scintigraphy, computer tomography scan, and nuclear magnetic resonance) represent the most powerful approaches in analysing the luminal transport, as some characteristics of flow patterns can be inferred from the observed wall motions. However, on a larger time scale, this kind of approach becomes very expensive and also harmful to the patient.

Each of these techniques has its own indications, advantages, and disadvantages. The most interesting approach for the evaluation of the intestinal motility is yielded by the Magnet Tracking device. This totally non–invasive investigation system, developed by the Institute of Physiology of the University of Lausanne, provides an estimate of the trajectory along the digestive tube of an ingested permanent magnet. In the next part of the dissertation, a description of the device is presented.

In this section, we are interested in creating an intestinal mass–displacement simulation from the simulated electrical activity and measuring the related motility patterns. So far, there are two major approaches in modeling the electromechanical coupling that occurs in the digestive tube and in the intestinal tissues composing it. The first approach uses dipole theory to describe the characteristics of the electric field produced by the excitable tissue. The dynamics of this field in the vicinity of this tissue represents its local electrical activity. Thus defining the relation between electric field and tissue contraction, this model permits to identify the mechanical response induced by a particular electric field excitation. Rashev et al. [62] use this type of model to test the effect of a microprocessor-controlled electrostimulation in the stomach [63] and in the colon [64]–[65].

The second approach consists in simulating the excitation–contraction coupling phenomenon of the smooth muscle with a model in which the internal contractile protein system mechanics

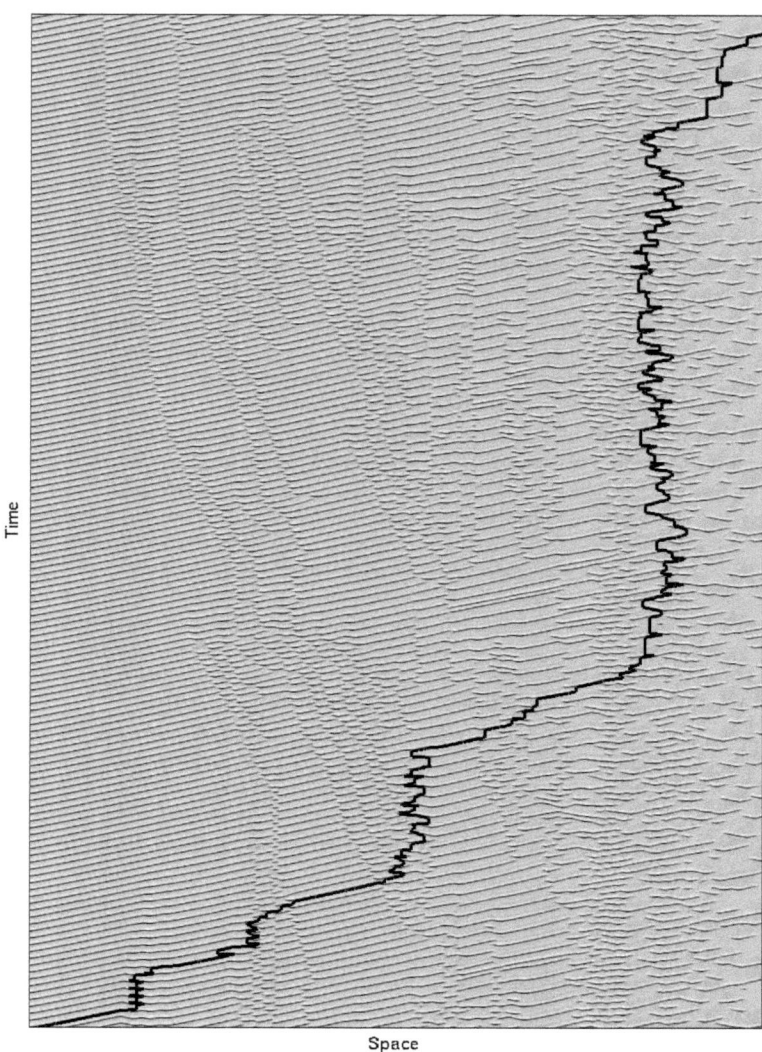

Figure 6.5: Space–time plot of intrinsic electrical activity obtained with the Aliev model. The black line indicates the trajectory of the virtual object drifted by slow waves, which are considered propulsion waves under the assumption that they are always accompanied by spikes.

is substituted by a relationship between the experimental force and calcium activity [66]–[67]:

$$T^a = \begin{cases} 0 & [Ca^{2+}] \leqslant 0.1\mu M \\ c_1 + c_2[Ca^{2+}]^4 + c_3[Ca^{2+}]^3 + c_4[Ca^{2+}]^2 + c_5[Ca^{2+}] & 0.1 < [Ca^{2+}] \leqslant 1.0\mu M \\ \max T^a & [Ca^{2+}] > 1.0\mu M \end{cases} \quad (6.4)$$

where T^a is the active force generated by the smooth muscle, and c_{1-5} are mechanical constants.

Both approaches are interesting, but of limited use and are not appropriate for the study of intrinsic motility. The approach we propose is founded on simple concepts, but provides stimulating results. Existing mathematical models of the gut employing a network of oscillators successfully describe the propagation of the slow waves. In the Aliev computer-model that we used for the simulations, only the longitudinal muscle layer is considered for intestinal slow wave simulation. It has been shown that the longitudinal and circular muscle layers of the distal colon contract and relax synchronously during propulsive movements [58].

Let us assume that slow waves are always accompanied with spikes. Each depolarisation wave in the longitudinal muscle layer of the model would thus be associated with a muscular contraction, and slow-wave propagation with propulsive movements. In order to see if slow waves are capable of displacing a virtual object through the intestine model, a point object characterized by a null mass is placed at the beginning of the model (oral part) and is allowed to drift with the slow waves to the end (distal part). The displacement direction of the virtual object is given by the direction of slow wave propagation and the velocity is modulated by slow-wave slope so as to have a stable and more realistic result. The figure 6.5 illustrates a typical simulation of mass displacement in the Aliev model.

Peristaltic activity allows intestinal contents to move in both directions, but preferentially analward. In the same manner, the trajectory of the moving object is characterized by forward and backward displacements, with periods of stagnation or rapid back and forth movements. The trail followed by the point object is determined by the slow waves and thus by the spatial and temporal organization of the whole system. It has to be noted that each trajectory is unique. By changing the initial time at which the object is released, the resulting trajectory changes.

6.5 Three–Dimensional Visualization

Three-dimensional (3D) visualization of human organs employing currently available medical imaging modalities, such as computed tomography (CT) and magnetic resonance imaging (MRI), is becoming an increasingly important technique in surgical planning, noninvasive diagnosis, and image-guided surgery.

This visualization technique is widely used for the 3D reconstruction of the brain and the heart, but the problem becomes harder when organs with no exact shape have to be represented as a precise volume, which is the case for small intestine. In fact, the gastro-intestinal tube is continually moving inside the abdominal wall and contracting itself, without any precise shape. Nevertheless, it is today possible to obtain a 3D reconstruction of some of its segments and also to exploit this powerful visualization technique to help physicians in their diagnosis and alleviate patient discomfort.

Currently, colonoscopy is the mostly used procedure for the detection of polyps or carcinoma of the colon. This widely used procedure exhibits unfortunately some uncomfortable drawbacks.

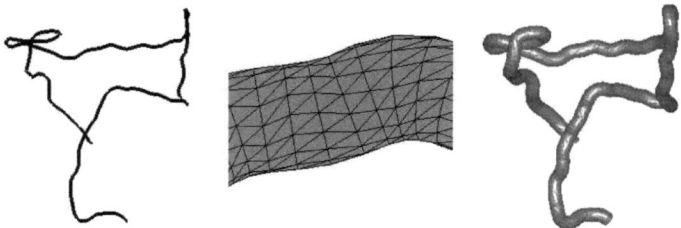

Figure 6.6: On the left, the line describing the 3D trajectory obtained from a virtual colonoscopy. In the middle, zoom on the triangular grid of the tube model. On the right, a tube model positioned and centered along the trajectory.

It is an invasive technique with a risk of perforation, requires intravenous sedation, and the proximal segment often cannot be fully evaluated because of technical difficulties and/or patient discomfort. New advances in this technology, still under development, permit to develop an alternative to optical colonoscopy. The virtual colonoscopy technique is a minimally invasive screening method that visualizes the inner mucosal surface of the colonic wall, and therefore assists the physicians in detecting the presence and characteristics of mucosal lesions [69].

Producing images of the mucosal surface of the colon involves several steps [70]. After the colon of a patient has been cleaned and distended with air, a spiral CT scanner is employed to obtain a sequence of thin axial slices from the top of the splenic flexure of the colon to the rectum based upon landmarks obtained from the scout image. Then, this set of CT images is reconstructed into a 3D volume and images of the inner structure of the colon can be visualized. This provides the capability to precisely explore the inner surface of the colon for irregularities such as polyps and tumours.

This 3D geometrical reconstructions are also widely used for computer modeling of organs, e.g. the geometrical model of the heart or a part of it, such as atria [31]–[34]. In this work, only a simple 3D reconstruction of the colon is used, in order to qualitatively show the results obtained with a model simulating the colonic electrical activity. This model, presented in the next part of the dissertation, is similar to the Aliev model of the small intestine, thus, it is a cable model generating monodimensional data that represent the electrical activity of a longitudinal section along the large intestine (see figure 4.10). The hypothesis of Aliev is that slow waves propagating in the longitudinal direction, can be considered as rings moving along the tube. Thus, in the model, the transmembrane potential of a circumferential section is represented as that of a single cell. Using the same hypothesis, the 3D representation of the colonic electrical activity is obtained by associating a ring to each point of the cable model, so as to obtain a tubular structure. The colonic shape is then obtained by positioning and centering the tube model along a 3D trajectory of a virtual colonoscopy. The axial line of our colon is represented on the left of figure 6.6, after a simple lowpass filter processing. At each point $p(i)$ of the 3D trajectory, a circle of radius R is associated and centered on $p(i)$, and finally located on a plane normal to the vector $p(i+1) - p(i-1)$. Note that the circle radius should not be greater than the minimal curvature radius along the trajectory, otherwise the circles will intersect. Along the circle, K points are equidistantly positioned, in order to create the nodes of the grid. A triangular grid is generated by binding the nodes as depicted in the middle of figure 6.6, while the final tube model of the colon is illustrated on the right of figure 6.6.

6.6 Conclusion

The numerical models provide a large amount of data, but have to be processed in order to permit a comparison with published experimental or clinical observations. In this chapter, some techniques are presented, in which results from the model are treated so as to confront them with real measurements. This operation definitely reduces the amount of available information, but is a necessary step toward model validation.

Space–time plots constitute a useful visualization technique used to analyse experimental measurements, in which electrical or mechanical activity of the intestine are recorded. It has been shown that this representation is also appropriate to visualize the behaviour of the oscillatory systems, such as the intestinal model.

The space organization of the different segments of intestine and its evolution is a basic but challenging question for researchers. That is why we propose a simple and robust tool for the temporal tracking of oscillator dynamics from the intestinal space–time plots. Our approach is based on the cross–covariance and can be used with simulated data as well as experimental observations. This new graphical tool has been illustrated on the Aliev computer–model.

The direct access to membrane potentials available in computer models allows us to estimate the extracellular potential and thus the electrograms of the intestinal tissue. Unipolar and bipolar measurements are modeled and discussed.

The mass–displacement is modeled using the simulated intestinal electrical activity and assuming that slow waves are always accompanied with spikes. Each depolarization wave can thus be associated with a muscular contraction and slow–waves propagation to propulsive movements. A point–shape object characterized by a null mass is then placed at the beginning of the model (oral part) and drifts with the slow waves to the end (distal part).

In the last section a simple 3D reconstruction of the colon is presented, the goal being a graphical visualization of the results obtained with a model simulating the colonic electrical activity.

Part III

Validation and Applications

The evaluation of the model is an important part of the modeling process. In order to recognize the importance of this delicate phase, the following question is suggested: does a computer model appropriately describe the system? We can assume that a computer model represents adequately the reality if its outputs are consistent with a set of system measurements. However, the question still remains open. In fact, measurement data have to be a representative set of possible values, and also: does the model describe events outside or between the measured data?

Chapter 7

Intestinal Electrical Activity

The smooth muscle of the gastro–intestinal tract is subject to an almost continuous but slow electrical activity. This activity tends to exhibit two basic types of electrical waves, namely slow waves (0–0.3 Hz) and spike bursts (2–4 Hz). Most gastro–intestinal contractions occur rhythmically, and this rhythm is determined mainly by the frequency of the slow waves. Slow waves themselves usually do not cause muscle contraction; instead, they mainly control the appearance of intermittent spikes potentials, which are the direct trigger of the contractions (except in the stomach).

The investigation of slow waves and spikes is the topic of intense research both from the experimental and simulation viewpoints so as to understand their functions and anomalies. Many experiments, in different ways, have brought information upon the electromechanical behaviour of small gastro–intestinal segments, but the activity of the entire gastro–intestinal tract is still not fully understood. Also, due to numerous technical problems, investigation of the spatial and temporal organization of the whole system is today only possible with the help of models.

This chapter is dedicated to the investigation of the electrical activity of the gastro–intestinal tract, from the experimental and computer model viewpoints. The first section presents the signals obtained from *in vivo* measurements, as well as the experimental setup. In the second section, the results of *in silico* experiments obtained with the Aliev computer-model are showed, discussed, and when possible compared to those of real experiments.

7.1 *In Vivo* Acute Experiments

The purpose of this section is to investigate the electrical activity of the gastro–intestinal tract. For this purpose, *in vivo* electrical rhythms were recorded from three different segments of the porcine tube: the stomach, the caecum, and the sigmoid. The electrodes were implanted by pairs in the serosa after a median laparotomy, providing bipolar recordings.

7.1.1 Preparation of Animals

Experiments were performed on 5 healthy adult farm pigs (2 males and 3 females) with an average weight of 41.2 ± 6.8 kg. All the animals were normally fed and had free access to water. They were operated under anaesthesia after overnight fasting.

Intramuscular Ketamine (10 mg/kg) premedication was used and general anaesthesia was

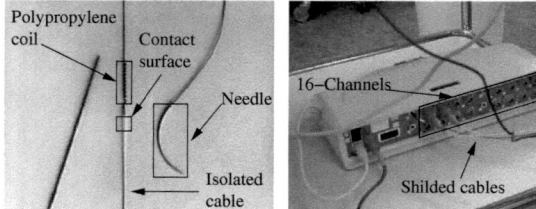

Figure 7.1: View of the Medtronic 6500 electrodes, on the left, and the backside of the Polygraf ID system on the right.

induced by inhalation of Isoflurane via a muzzle mask. A cuffed endotracheal tube, with the inner diameter of 9 mm, was then passed and the lungs were ventilated with a mixture of air, oxygen, and Isoflurane, for the maintenance of anaesthesia. Oesophageal temperature was measured and maintained at 38 ± 0.5 °C by a heating element incorporated in the operating table. Animals were constantly perfused with Hartmann solution at a rate of 3–5 ml/kg/h and their arterial blood pressure was monitored from a cannula inserted into the carotid.

A median laparotomy was performed under general anaesthesia, so as to expose the small intestine and the colon. A pair of electrodes was implanted in the muscular surface of the gastro-intestinal tube at different locations, such as the stomach, the caecum, and the sigmoid. At the end of the recording session, animals were sacrificed with an overdose of barbiturate.

7.1.2 Description of the Equipment

The electrodes implanted for the electrical activity recordings are the Medtronic 6500 Unipolar Temporary Myocardial Pacing Leads. Differential recordings were performed with a distance between the two electrodes of about 1 cm. For larger distances the recorded signal became too noisy. These electrodes are made of stainless steel and its contact surface is 2 mm long. It ends with a needle, which allows us to install the electrode in the serosa, and a polypropylene coil, permitting its fixation in the intestinal wall. The electrode ends on the other side with an insulated cable, which is connected to the acquisition system using coaxial shielded cables.

Electrical activity was recorded using the amplification–sampling system Polygraf ID (Universal Amplifier Version 1.1, Medtronic Functional Diagnostics in Denmark), connected to a laptop via USB cables. The observed electrical signals were written in files in ASCII format. The recording system is designed for animal use only. It can record up to 16 single or differential channels that have to be calibrated before the experiments. Its amplification gain can be programmed to 1, 10 or 100 and the sampling rate can be of 105 Hz, 209 Hz or 1674 Hz. Results presented in this work have been recorded using an amplification gain of 10 and a sampling rate of 105 Hz.

7.1.3 Results

Several recordings of electrical intrinsic activity have been performed. Figure 7.2 shows examples of signals recorded in the stomach, in the caecum, and in the sigmoid during *in vivo* experiments.

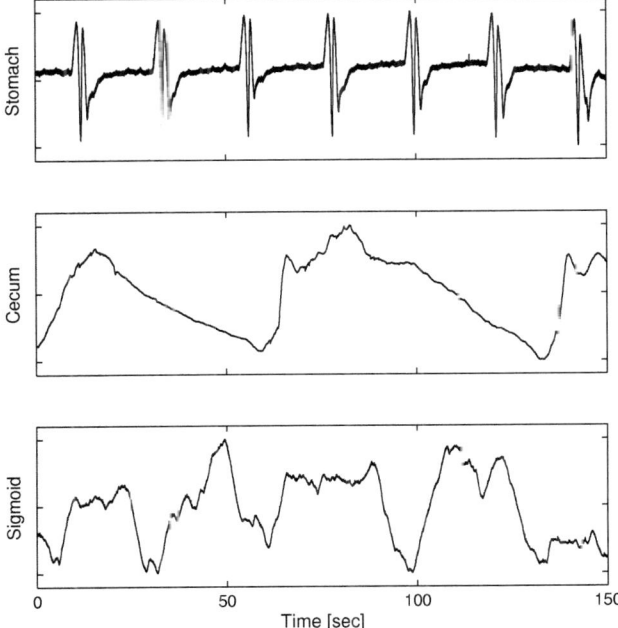

Figure 7.2: Intrinsic electrical activity of three porcine gastro–intestinal segments: the stomach, the caecum, and the sigmoid.

The first remark concerns the waveform of the signal events. They are recorded with a differential configuration of the electrodes, thus their shapes are also determined by external factors, e.g. the distance between the recording electrodes, the types of perturbation, the distance from noise sources, etc.

As the name suggests, the slow waves manifest themselves by slow cycles of depolarization and repolarization of the membrane potential. The effect of these slow cycles appears in the serosal electrical recordings with very low frequencies, which implies long recording periods to obtain some frequency information. Unfortunately, the type of anaesthesia used in these *in vivo* experiments drastically reduced the intrinsic gastro–intestinal activity. About 3 hours after the beginning of its employment, no more electrical activity was observable.

The recorded signals are perturbed by noise that can come from internal and external sources. The internal noise is generated by physiological activities, such as respiration and heart beating. This type of perturbation can be harmful if its frequency spectrum superimposes with the spectrum of the signal of interest. Electromagnetic external perturbation sources are generally easily filtered. In contrast, external noise generated by human activity, such as manipulations of the electrodes or the animal, gives rise to a dramatic degradation of the signal. Moreover, several minutes are needed until the recovery of normal behaviour. The only solution to this inconvenience is to stay absolutely still during electrical recordings.

Electrical activity of the stomach has a very regular frequency of about 3 bpm. Gastric signals are effortlessly recordable, which may be due to the persistence of the slow waves and its stability in frequency, and/or to the amplitude of the electrical activity giving rise to advantageous SNR. The intrinsic activities recorded on the caecum and the sigmoid are much more irregular with a frequency of about 1 and 2 bpm respectively. Actually, colonic slow waves are more difficult to record and do not have always a precise frequency. In fact, the signal consists of a repetition of electrical events that are often arduous to identify. Moreover, the recorded signals can exhibit periodic repetition of events, sporadic appearance of events, or nothing at all. The few measurements in the large intestine are often not reliable and do not permit a precise and unique frequency mapping.

The problems listed above show the fragile equilibrium governing this kind of measures, which can easily disappear or become too noisy. The technical difficulties added to the invasiveness of the experiment, fix the limits of this observation method at the experimental stage.

7.2 Simulation of the Electrical Activity Using Computer Models

In this section, we compare the results obtained using three different frequency gradients of the intestine model of Aliev. The only difference between the three models lies in the distribution of the intrinsic frequencies of the oscillators along the ICC layers. The first situation corresponds to a linear frequency distribution from 20 bpm at the beginning of the ICC layer to 8 bpm at the end. The second model is characterized by an exponential frequency distribution, whose fitting was performed on experimental values of the small intestine by Aliev. We used a similar approach (the third situation presented) to obtain a model of the large intestine by adjusting the parameter $\varepsilon_{\text{ICC}}(i)$ in order to mimic the experimental observations of the myoelectrical activity measured in the human large intestine [51]. For the small and large intestine models we used a one–dimensional array composed of $N = 300$ coupled cells for each layer. Comparing the array to a human intestine, the spatial discretization corresponds to a section of 0.8 cm per cell and 0.3 cm per cell for the small and the large intestine respectively.

Figure 7.3 shows the respective distributions of intrinsic frequencies for the three simulated models. Thick lines represent the intrinsic frequencies along the LM layers for a system without coupling between cells ($D_{\text{LM}} = 0$ and $D_{\text{ICC}} = 0$), while thin lines show the measured frequencies during computer simulation of the coupled system. Note that the measured frequencies are not in accordance with the intrinsic frequency of the oscillators. These results reflect the experimental observation that the measured distribution of intrinsic frequencies in an intestine, which is normally measured on tissue strips, does not coincide with *in vivo* experimental measures [71].

The presence of a sufficiently strong coupling between oscillators leads to local frequencies of excited collective oscillations that strongly differ from the natural frequencies. Given sufficiently large coupling coefficients, the oscillators are synchronized all though the array, i.e. global synchronization occurs. On the opposite, the dynamics of the oscillators become independent when zero coupling is used [72]. Intermediate values of coupling give rise to the formation of characteristic and well studied structures, namely clusters. These aggregations of neighbor oscillators with synchronized dynamics delimit the regions along which waves can propagate without interruption or disturbances. Clusters define also a propagation distance of slow waves

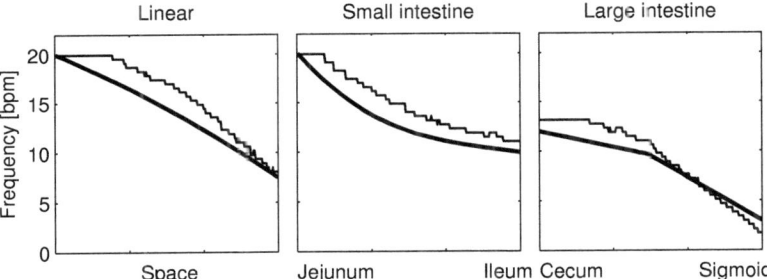

Figure 7.3: Intrinsic frequency of oscillators along LM layer for a system without inter-cell coupling (thick line) compared to measured frequency of oscillators for the coupled system (thin line): linear frequency gradient, small intestine frequency gradient, large intestine frequency gradient.

and are recognizable in figure 7.3 as plateau segments. The coupling strength regulates the extension of the clusters, i.e. the larger its value and the larger the cluster size. However, it is not the only factor influencing the clustering phenomenon; the frequency difference between two neighboring oscillators is also important. Consider two synchronized oscillators, and let us increase the frequency difference until they become no more synchronized. Thereafter, a stronger coupling factor is needed to reach the initial synchronized situation.

In our case, the two factors determining the clustering of the arrays are strictly bound in a relation established by the local value of the frequency gradient. The coupling factor is set to be homogeneous in each layer and its value equal for each model. Actually, the local value of the frequency gradient principally decides the clustering of the system. One can imagine to simulate the inverse situation, and thus to set the intrinsic frequency of oscillators constant along the ICC layer, and impose a coupling factor gradient. Simulations of this kind have been executed, and two major behaviours have resulted. If the system is run with identical initial conditions for every oscillator, a regime of global synchronization governs it. If random initial conditions are used, no propagation of waves has been observed in the space–time plots.

Figure 7.4 presents the space–time plots obtained with the computer models defined by the three frequency distributions illustrated in the previous figure 7.3. First of all, note that waves propagate in the three models along both directions. Forward waves propagate from the oral end (left) to the aboral end (right), and backward waves in the opposite direction. It is easy to remark that the three presented situations privilege the forward propagation direction. This is due to the fact that, under precise conditions, waves propagate essentially from the oscillator with highest intrinsic frequency to the one with lowest frequency. Backward propagations can also be generated and propagate along the models, but they are normally characterized by a shorter propagation distance. The locations where waves are generated and start propagating in both directions are called pacemaker zones, because of the correspondence of the phenomenon with the heartbeat one. It is interesting to observe that sometimes waves propagate without distortion and sometimes are perturbed or broken by other waves.

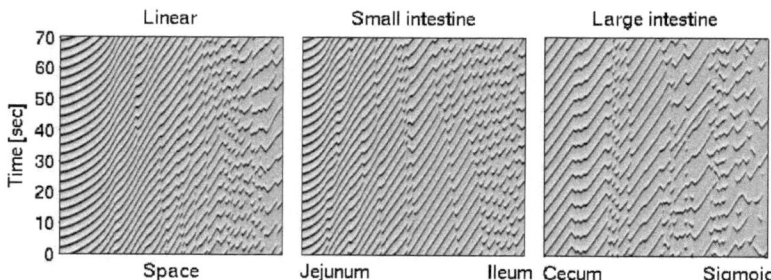

Figure 7.4: Examples of space–time plots of the transmembrane potential in the LM layer: linear frequency gradient, small intestine frequency gradient, and large intestine frequency gradient. Black and white correspond respectively to the maximum and the minimum values. Any propagating wave appears as an oblique black line.

So far, only the particularities common to the three situations have been commented. Therefore, each model exhibits a particular behaviour, which is distinctive and governed by the frequency gradient. Parameters as the oscillatory frequency, as well as the propagation distance and velocity of waves are the immediate and direct effects of the spatial and temporal organization of the systems. The patterns of space–time plots are thus different for each of the three models, and also within the same model if two distinct time windows are considered. In fact, these patterns change with time (in a manner similar to that observed with experimental data) due to the intermittent nature of the self–organization of the system.

Intestinal space organization evolves slowly, and it takes a long time to witness meaningful changes. Figure 7.4, however, shows the system dynamics for only a short time period (70 s) obtained with the Aliev model. This is a major limitation of such space–time plots. With the method we proposed in the previous chapter, i.e. the cluster evolution analysis, we can obtain a significant information compression of space–time organization, allowing us to observe the evolution on a one–hour scale in a single graph.

Observation of figure 7.5, on the left, inspires some remarks. First of all, there is a stable cluster at the beginning of the array. It is followed by an extended zone of intermittent organization, where black spots indicate the presence of large clusters. The last segment of this imaginary intestinal tract with linear frequency gradient is characterized by a decrease of the system organization.

For the small intestine, we can observe in the middle of figure 7.5 that a small stable cluster is present at the beginning of the system, which represents a group of functional units with related dynamics. A zone of periodic organization follows it. The central part shows a typical pattern of irregular cluster sizes, which become larger at the end of the system.

Figure 7.5, on the right, illustrates organization for the large intestine. One recognizes three main regions along the colon segment. The first one contains a stable cluster with variable dimensions, while the second one consists of clusters of weak spatial organization. The third one is characterized by an increased disorganization. These results confirm some experimental

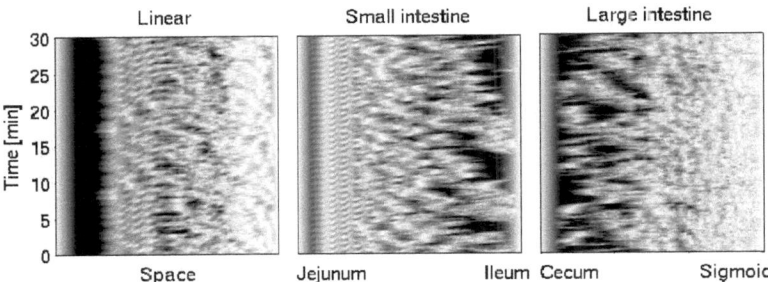

Figure 7.5: Evolution of the cluster size: linear frequency gradient, small intestine, and large intestine (from left to right). Black corresponds to a value larger than 60 cells, white to zero.

observations [13].

From the spatio-temporal plots, electrograms can also be computed using a transmembrane current dipole model. Transmembrane currents are first computed, and electromagnetic fields are then derived and averaged for 24 virtual surface recording electrodes that compose an array electrode. The 24 signals are obtained from sites located along the model, corresponding to the longitudinal axis of the organ. Figure 7.6 illustrates the electrograms provided by the virtual array-electrode for two patterns of slow-wave propagation. Both figures are obtained by positioning the array electrode at 8 mm from the tissue surface and considering it is 8 cm long. The slow waves consist of a negative moving deflection that slowly returns to the baseline. To visualize slow-wave propagation, signals provided by the 24 serosal-surface electrodes are displayed on top of each other. On the top of figure 7.6, a slow wave propagation pattern is illustrated. In this particular case, the waves spread in the aboral direction without interference as indicated by the arrow. However, the propagation pattern is not always stable and, as showed on the bottom of figure 7.6, pacemaker can start firing spontaneously at any time. These pacemakers then modified the pattern of propagation in the tissue, generating backward wavefront colliding against wavefronts propagating in the opposite direction.

These results also describe the patterns of slow wave propagations and are extremely similar to the electrograms recorded by Lammers in its experiments. Using an array electrode consisting of 240 surface electrodes (10 rows of 24 surfaces electrodes), he obtained a mapping of the small intestine of rabbits [54] and cats [55]-[56].

7.3 Conclusion

In this chapter, the electrical activity of the gastro-intestinal tract is investigated from the experimental and the computer model viewpoints.

Differential recordings are effectuated *in vivo*, implanting the Medtronic 6500 electrodes in the serosa of three different segments of the porcine tube: the stomach, the caecum, and the sigmoid. The stomach shows a very regular electrical activity, while the caecum and the sigmoid exhibit a sporadic and irregular behaviour.

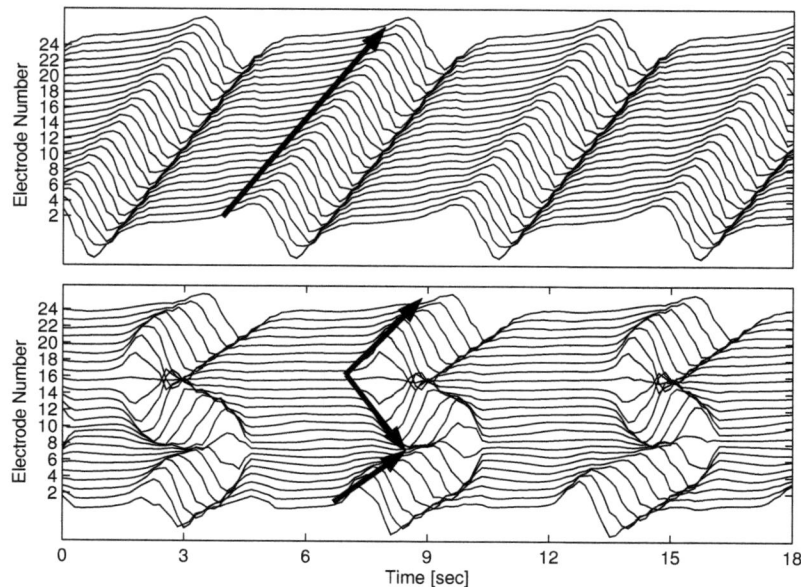

Figure 7.6: Simulated electrograms with an electrode array of 8 cm length, located at a distance $d = 8$ mm from the tissue surface. On the top of the figure, slow waves propagate without wave breaks. On the bottom, a pacemaker generates slow waves that stop propagating in caudal direction after breaking against slow waves propagating orally.

Three versions of the Aliev computer–model are presented. The difference between these three models lie in the frequency gradient along the ICC layers. Space–time plots are used to illustrate the electrical activity of the systems and allow us to extract interesting information, such as the slow–wave propagation velocity and distance and temporal pacemaker locations. Cluster evolution analysis, which is the method presented in the previous chapter for the evaluation of the space–time organization, is then used to compare the self–organization of the three systems. From the space–time plots, electrograms are also computed using a dipole model. Transmembrane currents are first computed, and electromagnetic fields are then derived and averaged to obtain the virtual surface–electrode recordings.

Chapter 8

Intestinal Mechanical Activity

Gastro–intestinal motility is generated by the contractions of the smooth muscular layers. The spatial and temporal patterns of the contractions in the gastro–intestinal tract are regulated by a complex interplay between the mechanical, neural, and chemical control mechanisms. These contractions appear in an organized manner along the entire tube and are responsible for the mixing and propulsive movements.

Beside the common techniques used to investigate intestinal motility, which are generally invasive and limited for research, promising methods based on magnetic measurements have been developed. *In vitro* experiments are also important, but do not permit to observe the motility of the entire intestinal tube and to investigate with fully physiological conditions.

In this chapter, the mechanical activity of the gastro–intestinal tract is investigated, both from the experimental and computer model viewpoint. The first section presents the signals obtained from *in vitro* measurements. In the second section we present some signals of *in vivo* mechanical activity, as well as the experimental setup used for the recordings. The last section is dedicated to *in silico* experiments.

8.1 *In Vitro* Experiments

Four *in vitro* experiments were performed in this study, with the objective of provide a better insight on the intrinsic activity of the colon. The experimental setup is presented in this section. The frequency and the amplitude of the contractions were measured directly from a colonic tissue strip. The spontaneous smooth muscle activity from the porcine sigmoid was recorded, and responses of this tissue to the some physiologically relevant agents, such as Dormicum, Atropine, Acetylcholine, were observed.

8.1.1 Description of the Equipment

The experimental system consists of several elements as illustrated in figure 8.1. The bath is filled with a physiological solution, oxygenated by a constant amount of air bubbling. Heated water at 37°C flows continuously in the warming circuit so as to maintain the temperature of the physiological solution constant in the bath. A longitudinal and a circular (transversal) strip are removed after sacrifice from the sigmoidal segment of a porcine intestine. One end of the intestinal strip is attached to a clip located on the base of the bath, and the other end to the

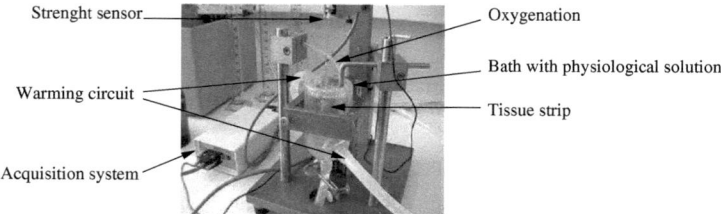

Figure 8.1: Experimental setup for *in vitro* experiments.

hook of the transducer. The tension on the tissue has to be adjusted to allow the sensor to record the signal. Spontaneous contraction of the intestine stretched to different lengths can be measured.

For recovery to normal function, three to five minutes are needed after the strip is fixed in the bath. Slow contraction waves through the strip should be clearly visible once normal function has been restored. The temperature of the warming circuit has to be continuously monitored, and the stream of bubbles has to be controlled to prevent the strip from being moved by the bubbles.

The effect of agents, such as Dormicum, Atropine, and Acetylcholine, are tested by adding a specific amount of them to the physiological solution in the bath. Once the effect on the tissue strip activity is observed, the bath fluid containing the agent is drained, carefully rinse the intestine preparation with fresh and warm (37°C) physiological solution a couple of times, and refill the chamber. This removes the agent from the bath and thus reduces its effect on the tissue.

8.1.2 Results

This experimental setup permits to directly measure the amplitude and the frequency of the spontaneous contractions of the smooth muscle. Figure 8.2 shows a typical recording of longitudinal and circular smooth muscle activity of the porcine colon. Both tissue strips exhibit rhythmical contractions at about 1 bpm. The spontaneous rhythm is easily perturbed by the use of drugs or the variation of physiological conditions. We tested the effect of some relevant drugs on the intestinal motility, in order to design an anaesthetic protocol for *in vivo* acute experiments on the observation of the intestinal activity. Atropine, Ketamine, and Dormicum show a dramatic effect on the intestinal motility. After being added in the bath solution, the spontaneous activity disappears and the signal becomes flat. These drugs have thus been omitted from the anaesthetic protocol presented in the next chapter, where the preservation of motility is fundamental to the evaluation of efficacy of the electrostimulations.

The variation of physiological conditions, such as tissue stretch or temperature, have also important effects on the motility, but are not discussed in this work.

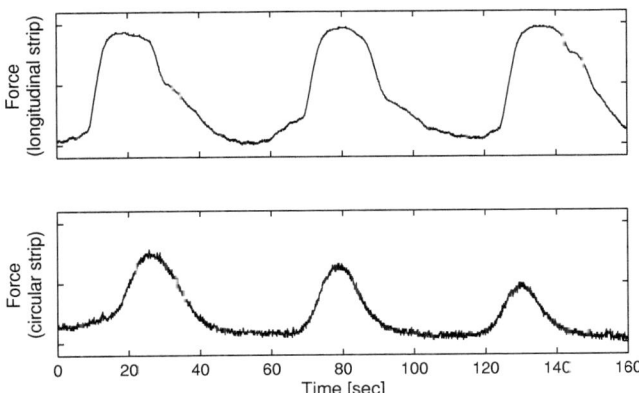

Figure 8.2: *In vitro* intrinsic activity of a longitudinal and circular strip of porcine sigmoidal tissue.

8.2 *In Vivo* Experiments

The objective of this section is to investigate the mechanical activity of the gastro–intestinal tract. For this purpose, a magnet is ingested by a pig and its trajectory in the gastro–intestinal tract is measured by Magnet Tracking. These *in vivo* experiments provide signals that contain interesting information, such as the rhythm of the intestinal contractions and the mass displacement velocity. A powerful signal processing technique for the information extraction from Magnet Tracking measurements is also presented.

8.2.1 Description of the Equipment

The Magnet Tracking, developed by the Institute of Physiology of the University of Lausanne, is a minimally invasive investigation system providing information about the movement of the luminal content [73]–[74]–[75]. It consists of an array of magnetic field sensors tracking a cylindrical permanent magnet of about 6 mm diameter and 7 mm length with a density between 1.3 and 2 g/cm^3 (figure 8.3). Dimensions and weight of the magnet are very important for technical (accuracy) and physiological (ingestion, natural progression throughout the intestinal tube) reasons. The sensing system is a 4×4 array of 16 cylindrical Hall sensors with integrated flux–concentrators, and is connected to a laptop via serial port. The signal of each sensor is amplified with a low–noise electric circuitry designed for this specific application. The position of the magnet is fully defined by 6 coordinates: 3 translations (X, Y, and Z) and 2 rotations (θ and ϕ). With a dipole, the rotation around the magnetization axis (symmetry axis) will not change the measured magnetic field. Translations and rotations of the magnet are recorded with a sampling frequency of 10 Hz and can be displayed in real–time on the laptop.

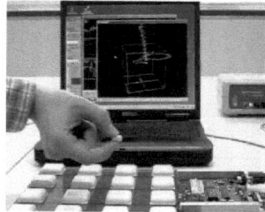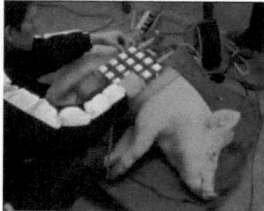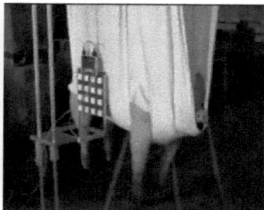

Figure 8.3: View of the Magnet Tracking system, on the left side, with its two main components: the magnet and the array of sensors. On the laptop screen is shown the 3D–trajectory of the magnet. In the centre and on the right side, Magnet Tracking experimental setup on an asleep and awake pig respectively.

8.2.2 Description of the Data Processing Methods

The signals recorded by the Magnet Tracking device contain several informations. These signals consist in the superposition of respiration movements, rhythmic intestinal contractions, and propulsive movements. Parts of the information is directly available, but some of it has to be extracted from signals using specific signal processing techniques. The best results are obtained using morphological filtering.

Mathematical morphology is a powerful concept for studying the geometrical properties of a signal [76], and has become particularly popular in image processing [77]–[78]–[79]. In this theory, the signal is viewed as a set in Euclidean space and transformed by several set–processing morphological operations. The basic operations are erosion, dilation, opening, and closing [80]. These transformations use a structuring element to interact with the set and extract information. A structuring element is a set of simpler size and shape and is chosen for its geometrical properties.

To understand how these operations transform a set X, an example of each operation is illustrated in figure 8.4. The structuring element in this case is a horizontal line with the origin at the centre of symmetry. On the left of figure 8.4, we can observe how erosion and dilation shrink and expand the signal. The opening transformation filters out the positive peaks, while the closing one filters out the negative peaks. It is important to notice that the order of operations is important because the latter are not commutable.

Morphological filters are nonlinear filters and consist of several morphological operations to be performed with various structuring elements. In our case, the morphological filter is constructed to filter respiration and to extract intestinal rhythms and magnet trajectory from Magnet Tracking signals. For this purpose, a precise combination of morphological operations is determined according to the characteristics of the signals and the properties of the basic transformations described above.

The opening of a closing and its dual are well–known operations in morphological filtering, giving rise to a smooth signal where positive and negative trajectory peaks, with a width smaller than that of the structuring element, have been suppressed. In order to suppress the small oscillations due to respiration, we compute a combination of the opening of a closing and its dual using the same structuring element. The sampling frequency of the Magnet Tracking signals is 10 Hz and the shape of the structuring element $SE_1 = 0.2\,[-1:1/25:1]^2$ has been empirically

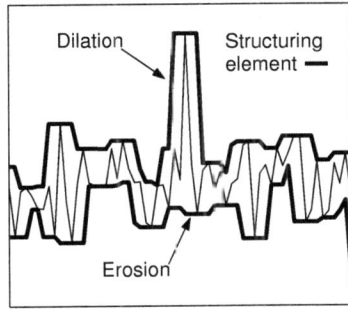 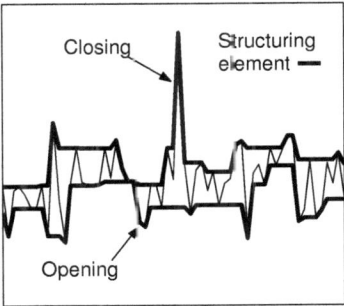

Figure 8.4: The four basic operations of the mathematical morphology: erosion and dilation (on the left), opening and closing (on the right). The structuring element used to filter the original signal (thin line) is a horizontal line and is also displayed.

determined by the signal characteristics. The resulting signal is clean from respiration and can be processed to separate intestinal rhythms from magnet trajectory A bigger structuring element $SE_2 = 0.1\,[-1:1/100:1]^2$ has to be used to extract the rhythmic slow–variations around the position of the magnet. The closing gives rise to a upper trajectory and the opening to the lower trajectory, which permits the extraction of the positive and negative peaks respectively, by a simple signal subtraction, also known as "top hat". The contraction frequency can be evaluated by considering the interval between two consecutive peaks.

A practical example is illustrated in figure 8.5, where the principal filtering steps are showed. The original signal is represented in the first part with a thin line. The oscillations due to the respiration are visible. In the same graph, the signal free of the respiration is illustrated with a thick line. The structuring element used in this morphological filter is also showed. From this signal, the magnet trajectory in the gastro–intestinal tract and the contraction rhythm can be separated. The upper trajectory is depicted in the second line, as well as the structuring element used. The simple subtraction of two signals, the trajectory and the signal without respiration, provide a sequence of events representing the intestinal contractions, as illustrated in the third line by the thin line. In order to eliminate the small variations in the signal, only contractions larger than a fixed threshold, which is empirically identified, are considered (thick line) and taken into account for frequency evaluation.

8.2.3 Results

Three experiments related to the mechanical activity measurements are presented. The intrinsic activity was first measured on an awake pig, and recorded signals were processed to obtain the instantaneous frequency of the rhythmic contractions. Then, we show the effect of a drug on the intestinal motility of an awake pig. The third measurement took place during an acute experiment, the magnet being placed in the colon of an anaesthetized pig. Synchronous recordings of the electrical and mechanical intrinsic activity have been performed.

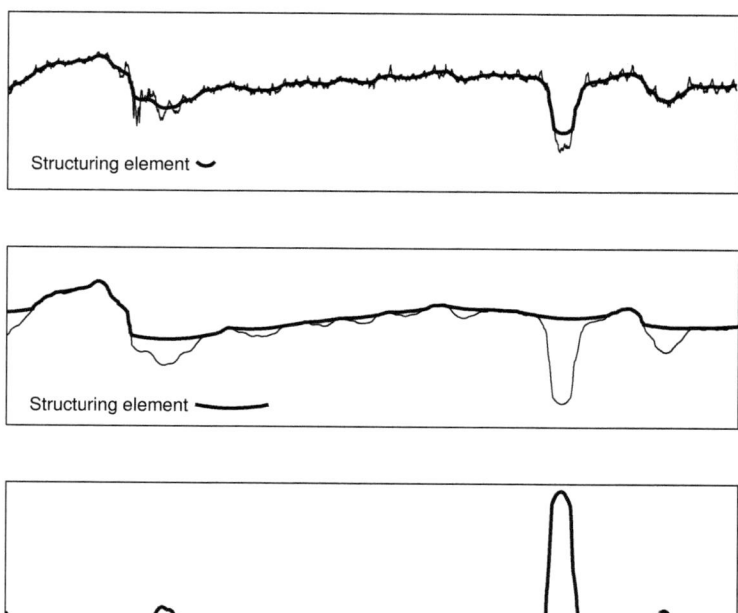

Figure 8.5: Procedure for signal separation. The original Magnet Tracking signal (thin line) and a respiration filtered version (thick line) is illustrated on the upper graph. The magnet trajectory is showed in the middle (thick line) and the intestinal rhythms are simply obtained by a signal subtraction, as depicted on the lower graph (thin line). A thresholded version of the signal (thick line) is considered to evaluate its instantaneous frequency.

Intrinsic Activity

Several applications and results of the Magnet Tracking system have been widely discussed in [75]. It permits to evaluate the trajectory of the magnet along the gastro–intestinal tube, as well as the transit time, the content propagation velocity, the length of the intestine, and the frequency of the rhythmic contractions.

Signals of the mechanical activity of the porcine colon, measured with the Magnet Tracking device, are shown in figure 8.6. The five signals record the movement of the magnet along the gastro–intestinal tube, that consists mainly of respiration, translations, and rhythmic movements. In this particular recording, the respiration is principally visible in the X, Y, and Z signals, and is recognizable by small and rapid magnet oscillations at about 15–25 bpm. The oscillations slower than the respiration, at about 3 to 12 bpm, are the rhythmic muscular contractions and are clearly visible on the θ and ϕ signals. The trajectory of the magnet displays effective movements

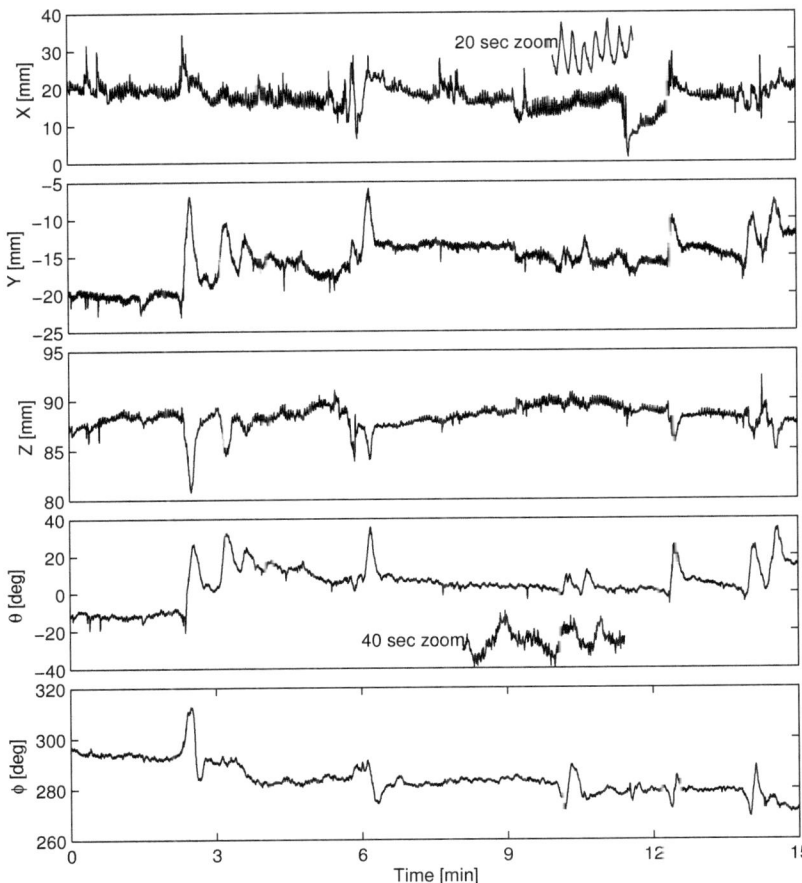

Figure 8.6: 15 min recording of the intrinsic mechanical activity of the colon. X, Y, and Z are related to the magnet position in 3D space, θ and ϕ to the orientation of the magnet.

along the intestine and can be well observed on the Y signal. The position at time $t = 0$ and at $t = 15$ min are different. The magnet moved of about 8 mm.

Signals recorded in the colon are generally difficult to analyse, compared to those of other intestinal segments. In fact, unlike in the upper part of the gastro-intestinal tract, it is rare to see regular rhythmic activity. Rhythmic contractions observed in the small intestine seem to lose their regularity in the colon, and classic methods of spectral analysis are no more efficient.

Colonic signals are therefore processed with respect to the scheme illustrated so as to extract

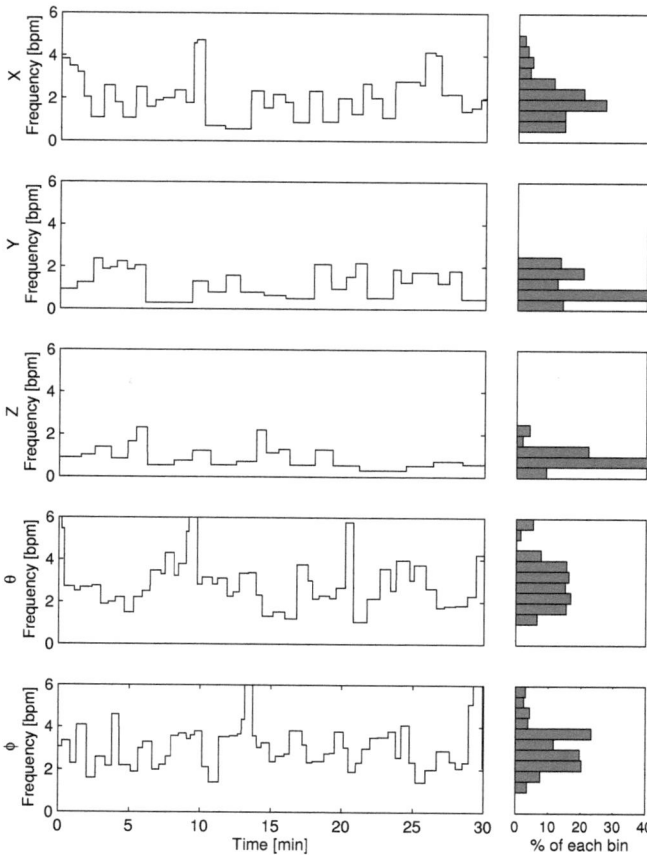

Figure 8.7: On the left the time evolution of the inverse of the contraction intervals are shown, and the histograms of their occurrences are plotted on the right.

the rhythmic contractions from the rest of the signal. The inverse of the intervals between successive rhythmic contractions, determining the instantaneous frequency, are plotted for the five signals on the left of figure 8.7. The histograms, on the right of figure, indicate the probability distribution of the corresponding instantaneous frequencies.

It is worth noting that Y and Z signals exhibit a probability distribution located around very low frequencies (0–2 bpm), while the probability distributions of θ and ϕ are translated to higher frequencies, about the double, and thus from 2 to 4 bpm. The frequency distribution for the signal X lies within these two situations, between 1 and 3 bpm. As observed in figure 8.6, rhythmic contractions are more visible, and easier to automatically detect, for θ and ϕ signals than for

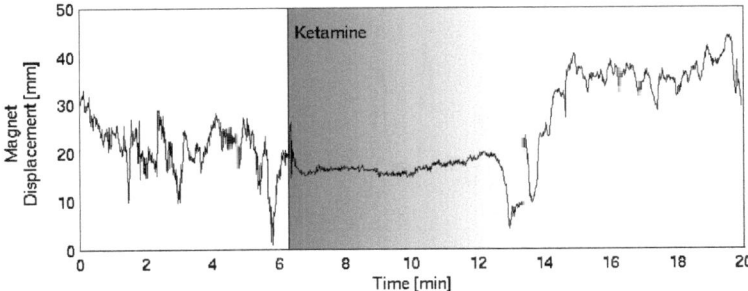

Figure 8.8: Effect of Ketamine anaesthesia on intrinsic mechanical activity of the caecum. The movement of the magnet inside the intestine before and after the injection of Ketamine is displayed. A significant reduction of activity after injection is observed.

the others. This explains the small translations of the probability distribution of rhythmic contractions. The frequency distribution for signals Y and Z show a mode corresponding to 0.5–1 bpm, and for X between 1.5 and 2 bpm (also visible in the signal Y) For ϕ a frequency of 3.5–4 bpm is the most common, and the signal θ does not show a precise mode. The signals of this recording thus show rhythmic contractions at about 2–4 bpm.

Effect of Drugs

In this second experiment we studied the effect of a drug, namely Ketamine, on gastro–intestinal motility. A magnetic capsule was swallowed by a pig some days before the measurements. The animal, that stayed at the farm, was placed in a hammock as shown on the right of figure 8.3. The pig rapidly got used to this situation, especially because its legs were hanging without touching the ground or any other support. It could stay still for at least one hour.

The magnet was localized in the colon and normal intrinsic activity was recorded during 20 min. In figure 8.8 is displayed the time course of a coordinate of the magnet position as well as the instant of drug injection. At the 6th minute of recording, a bolus of Ketamine was intramuscularly injected and immediately a drastic reduction of colonic activity was observed. At the 13th minute, thus 7–8 min later, the intrinsic mechanical activity restarted gently. This result confirms the observations of the *in vitro* measurements, where Ketamine also slowed down the mechanical activity of the colonic tissue strip.

Simultaneous Recordings

Synchronous recordings of electrical and mechanical intrinsic activity have been performed in the porcine sigmoid colon. The animals were prepared as explained in the previous chapter. A pair of Medtronic 6500 electrodes were sewed in the colonic serosa and a magnet was inserted at the same location as the electrodes. Figure 8.9 illustrates the recorded signals. We can observe a strong correlation between these two activities, with a frequency of oscillations of about 1 bpm.

The recorded mechanical movements are correlated to the electrical activity, yet the presence

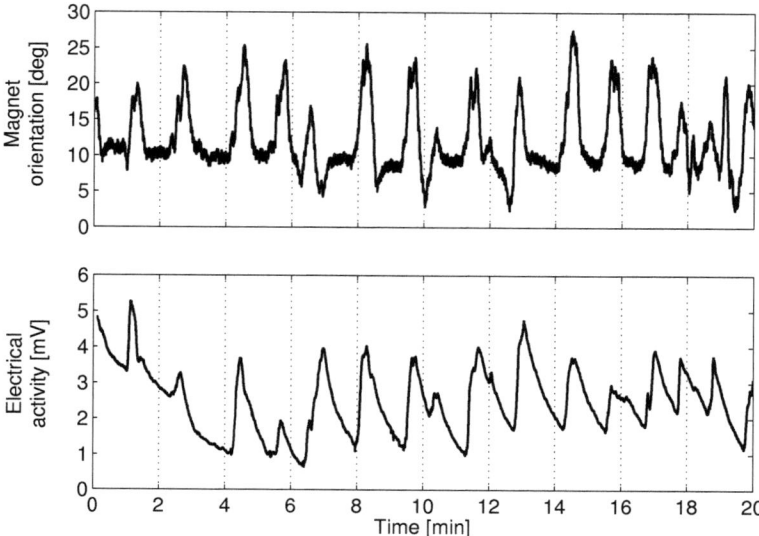

Figure 8.9: Simultaneous recordings of the intrinsic mechanical (top) and electrical (bottom) activities in the pig's sigmoid colon. Mechanical movements are shown as rotation angles of the magnet inside the caecum.

of an electrical activity does not necessarily lead to a mechanical contraction. In fact, studies on colonic physiology showed that mechanical activity is only observed when electrical activity is composed of spike bursts on top of the slow waves.

8.3 Simulation of the Mechanical Activity Using Computer Models

In order to see if a slow wave is capable of displacing a mass from the beginning to the end of the intestine model, we considered a virtual object to be dragged by the slow waves, and placed it at the beginning of the model, as described in chapter 6. By changing the initial time at which the object is released, the resulting space–time trajectory changes. Figure 8.10 illustrates the trajectories of three mass–displacement simulations, obtained using the same large intestine Aliev–model. We can see that the trajectories of the moving object are characterized by forward displacements, periods of stagnation, and backward and forward movements. The first part of the dotted trajectory is characterized by small forward and backward movements, during about one third of the transit time. This particular behaviour generates oscillatory movements and is very similar to the rhythmic contractions measured by the Magnet Tracking. Then, the trajectory shows a forward displacement of the object, with a slowing down of its propagation velocity when it approaches the aboral end. The dashed trajectory is characterized by forward

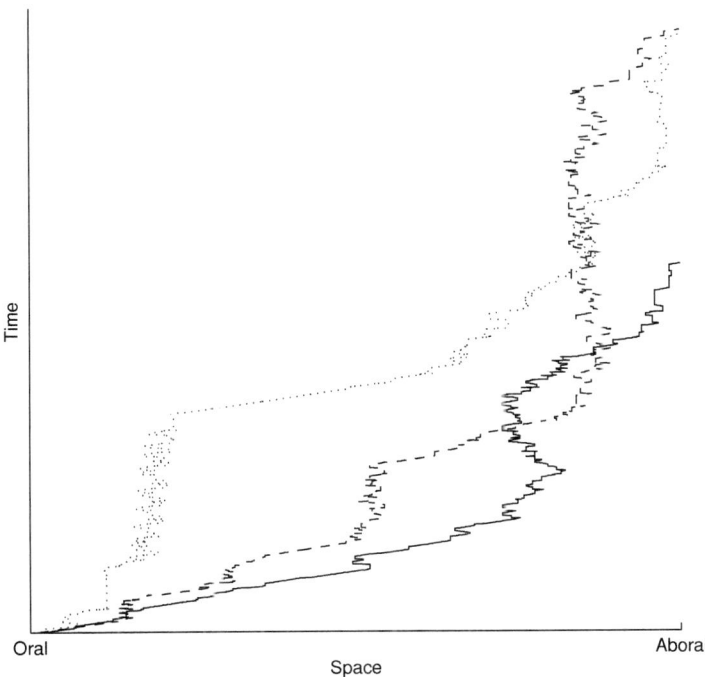

Figure 8.10: Three space–time trajectories obtained with the mass–displacement model. Each line represents a simulation with different starting time.

displacements along a segment, separated by a certain time interval in which the object exhibits little back and forth movements. The object stagnates one half of the total transit time. The last trajectory indicates a constant forward displacement along almost the entire segment, and a sudden backward displacement before reaching the aboral end.

The variability of the transit time, i.e. the total time an object takes to move from one end to the other is worth noting. Statistics of the transit time can be evaluated, but they do not fully reflect reality since slow waves do not always generate contractions, while it is the case in simulations.

Since each mass displacement simulation exhibits a different space–time trajectory, one can simulate a large number of trajectories and evaluate the probability distribution of the object location along the tube, as shown in figure 8.11.

The three cases exposed in this figure, correspond to the three different Aliev models presented in the previous chapter. The situation on the left of the figure correspond to the Aliev model with linear frequency distribution along the ICC array. It exhibits a constant increase of the probability distribution along the oral–aboral direction, meaning that the object is on average

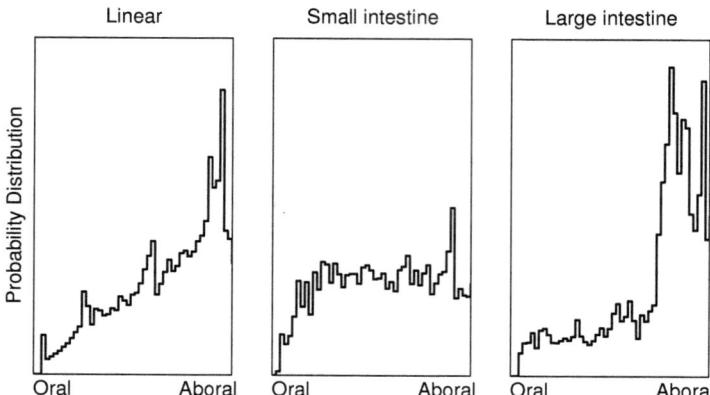

Figure 8.11: Probability distribution of the space–time trajectories simulated with three Aliev models: linear gradient, small intestine, and large intestine.

dragged more slowly as it approaches the aboral end. The four peaks interrupting the linear trend of the probability distribution correspond to the borders of the three organized regions identified on figure 7.5. In fact, the object stagnates in the intra–cluster zones, where organization discontinuities are observed.

The second simulation, corresponding to the Aliev model of the small intestine, exhibits a probability distribution that is uniform along the entire oral–aboral segment, except for the first oral segment that is not physiologically realistic as discussed in a previous chapter.

The probability distribution obtained with the simulations of the large intestine Aliev–model shows mainly two distinct regions. The first segment, the three quarters of the total length, displays a uniform distribution, and the last part a sudden increase, meaning that the object slows down and probably stop its progression when it approaches the aboral end.

The intrinsic electrical activity of the colon can be visualized on the 3D geometrical model presented in a previous chapter, as shown in figure 8.12. The simulation results are visualized using an arbitrary gray colour–map code, gray corresponding to the lowest value of electrical potential and white to the highest value that corresponds to the depolarized state. One can wish to display simultaneously in this visualization the result of the corresponding mass displacement simulation. A sphere is used to represent the position of the object that is dragged by the slow waves. Nine frames of the same sequence have been selected at regular time intervals. The object is, at the beginning of the simulation, positioned in the caecal segment, and at the end of the simulation it reaches the opposite end of the geometrical model. One can note that the first four frames of the sequence indicate a forward displacement of the sphere, while between four and six, we can observe a backward movement. In the same manner, the Magnet Tracking signals show the position of the magnet in the gastro–intestinal tube. Their time evolution during the simulation is displayed below each frame. The sphere displacement is given by the dynamics of the object as described above and the X, Y, and Z coordinates of its location are computed

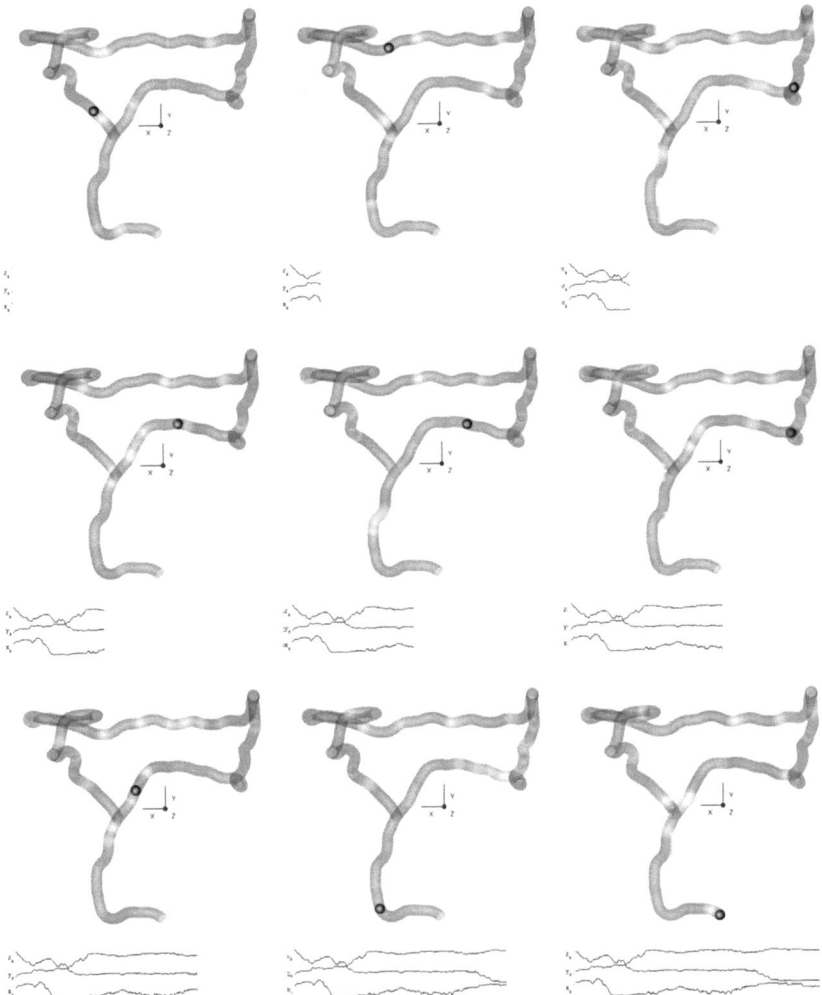

Figure 8.12: Simulation of mass displacement in the colonic computer model. The electrical activity is represented on the 3D geometrical model using an arbitrary gray-map code, white for the highest value and gray for the lowest. The sphere represents the position of the magnet. The simulated Magnet Tracking signals are illustrated below each picture.

from an arbitrarily chosen reference, as depicted below each frame. The time evolution of these coordinates can be compared to the time traces obtained by the Magnet Tracking (X, Y, and Z coordinates and 2 rotation angles for the Magnet Tracking). Similar patterns are also seen in experimental measurements using the Magnet Tracking system [75].

8.4 Conclusion

In vitro experiments presented in this chapter allowed us to observe the spontaneous circular and longitudinal smooth muscle activity of a colonic porcine strip, and evaluate the effect of some physiologically relevant agents on the mechanical activity of the tissue.

The assessment of the mechanical activity of the entire gastro–intestinal tract was performed in *in vivo* experiments using the Magnet Tracking. A magnet was ingested by a pig and its trajectory was tracked from the mouth to the anus. A useful signal processing technique based on morphological filtering for the extraction of instantaneous frequency of the rhythmic contractions from Magnet Tracking signals is presented and the results discussed.

The effect of the Ketamine on the intestinal motility of an awake pig was recorded, showing a dramatic reduction the mechanical intrinsic activity. Results also confirm the observations of *in vitro* experiments.

Synchronous recordings of the electrical and mechanical intrinsic activity of the porcine sigmoid in *in vivo* acute experiments have been performed. Signals indicate a strong correlation between this two phenomena, yet the presence of an electrical activity alone does not necessarily lead to a mechanical contraction.

In parallel to animal experimental work, we used the computer model of Aliev to reproduce experimental observations obtained with the Magnet Tracking system. Three different trajectories resulting from mass movement simulations have been discussed. We also simulated a large number of trajectories and evaluated the probability distribution of the location of the drifting object along the intestine. A 3D visualization of a geometrical model of the colon was used to display simultaneously the intrinsic electrical activity and the respective mass displacement simulation.

Chapter 9

Electrical Stimulation

The title of this chapter refers to the technique of electrical stimulation therapy for intestinal disorders, that is, in inducing peristaltic activity of the colon, so as to speed up the transit of intestinal content in this segment. Motility disorders, such as chronic constipation, constitutes difficult medical problems with a high treatment costs. Chronic constipation correspond to a decreased number of bowel movements and/or difficulties in evacuating the rectum, including excessive straining or hard stools. The resulting excessive slow transit is for many individuals the source of an extreme discomfort that can be caused by many factors, such as diet, insufficient physical activity, neurological or congenital disease, etc. Conventional medical solutions involve laxatives, diet [81] or surgery [82]. However, these therapies address mostly only symptoms and surgery is still based on empirical considerations, with sometimes disappointing results.

Indeed, the factors regulating the gastro–intestinal motility are not fully understood. Increase of motility is observed after eating [83], but the neuro–humoral mechanism through which the modulation is elicited is complex and not completely elucidated. Pharmacological stimulation and their secondary effects have been widely studied [84], though showing strict limitations on their long–term use. The abdominal wall massage is also a candidate for the treatment of chronic constipation, but its effect on the intestinal motility has not been completely acknowledged [85]. Promising results have been obtained with sacral nerve stimulation [86], successfully used in the treatment of urologic disorders and fecal incontinence. Some of the patients following the treatment and suffering from concurrent constipation have also noted improved stool frequency and rectal evacuation [87], but further studies are required.

Electrical stimulation could represent an alternative approach. Several studies of acute and chronic electrical stimulation have been performed over the last decade on different segments of the gastro–intestinal tract. Electrical stimulation of the stomach [88]–[89] is now accepted for the treatment of gastro–paresis with stimulation at a frequency of about 4–5 times the intrinsic basal rate [90]–[91].

Following a similar approach, different electrical stimulation settings have been suggested for generating slow waves in the small intestine. Lin et al. showed in nine dogs that it is possible to simulate electrically the small intestine with intraluminal electrodes to entrain slow waves [92]. Mosse et al. showed in twelve pigs that electrical stimulation of the small intestine could propel an endoscope by initiating local contraction [93]. Hughes et al. electrically stimulated the intestinal wall in seven dogs to produce contractions generating intraluminal pressure changes that led to an evacuation of colonic pouch [94].

Table 9.1: Published optimal parameters of electrical stimulation of the intestine. IF indicates intrinsic frequency.

Research group	Experimental model	Intestinal location	Stimulation parameters
Moritz et al. [101]	7 dogs	Jejunum	910 Hz, 0.2–0.5 ms, 25–30 mA
Hughes et al. [94]	7 dogs	Rectum	10 Hz, 0.5 ms, 15–20 V
Maw et al. [95]	9 pigs	Ileum and caecum	10 Hz, 2 ms, 100 mA during 2 min
Bruninga et al. [96]	5 cats	Rectum	40 bpm, 1 ms, up to 50 mA
Lin et al. [92]	9 dogs	Jejunum	1.1 IF, 70 ms, 4 mA during 25 s
Mosse et al. [93]	7 pigs	Small intestine	15 Hz, 30 ms, 12–20 mA
Amaris et al. [97]	5 dogs	Descending colon	50 Hz, 10 ms, 20 V during 18 s
Shafik et al. [100]	24 humans	Rectum	1.15 IF, 200 ms, 5 mA

While many experiments have been reported on the generation of slow waves via electrical stimulation of the stomach or the small intestine, few attempts of direct stimulation of the colon have been performed. Maw at al. stimulated electrically the colon in nine anaesthetized pigs and demonstrated a reduction of caecal volume measured via intraluminal pressure [95]. Bruninga also tested direct electrical stimulation of the colon in five cats leading to improvement of colonic transit [96]. In 2002, Amaris et al. stimulated transected descending colon in six dogs showing that sequential electrical stimulation accelerated significantly the movement of solid content [97]. Shafik et al. reported preliminary results of human colon stimulation: colonic pacing was able to initiate activity in patients with total colonic inertia, while in healthy volunteers an increase in frequency, amplitude and velocity of basal electrical waves was observed [98]–[99]–[100]. All these previous studies indicate that the use of an implantable stimulator for the colon can be a very promising approach. However, published optimal parameters cover a broad range: 0.05–1600 Hz for the frequency, 0.5–200 ms for the pulse width and 1–100 mA for the current intensity. All the stimulations were bipolar using several pairs of electrodes, either intraluminal or implanted surgically in the serosa. Table 9.1 summarizes the optimal pacing parameters for the techniques described. Although these parameters are in the same order of magnitude, there are some variabilities between the different protocols.

Several patents have also been issued in the field of gastric, intestine or colon stimulation, with different stimulation patterns:

- **US 6,104,955** "Method and apparatus for the electrical stimulation of the gastro–intestinal tract", I. Bourgois, 2000. Implantable pulse generator for electrical stimulation for patients having dysfunctional gastro–intestinal muscle or disorders of smooth muscle elsewhere in the body, interpulse interval 6–600 ms, amplitude 1–50 mA, pulse width 0.003–1 ms.
- **US 6,238,423** "Apparatus and method for treating chronic constipation", G. H. Bardy, 2001. Implanted stimulus generator with electrical stimuli provided to nerves of the autonomic nervous system or directly to the muscles from oesophagus to anus.
- **US 6,243,607** "Gastro–intestinal electrical pacemaker", M. P. Mintchev and K. L. Bowes, 2001. Electrical stimulation of smooth muscle with a timing mechanism to stimulate electrodes

successively and repetitively. Preferred embodiment: 50 Hz, 14–15 Vpp, 2–4 cm between electrode sets.

US 6,449,511 "Gastro-intestinal electrical stimulator having a variable electrical stimulus", M. P. Mintchev and K. L. Bowes, 2002. Algorithm described by Amaris in [97]. Local contractions with circumferential electrode sets artificially propagated by phase locking and time shifting of electrical stimulus.

US 6,658,297 "Method and apparatus for control of bowel function", G. E. Loeb, 2003. Electronic device implanted on or around the large bowel to stimulate peristaltic contractions.

WO 97/31679 "Biological electrostimulator of viscera", V. N. Dirin and A. G. Martusevich, 1997. Ingestible ovoidal electronic device stimulating with electrical pulses the luminal surface during its trajectory along the gastro-intestinal tube.

The first section of this chapter is dedicated to the presentation of the CTI project number 4921.1, whose results are showed and discussed in the following sections. The goal of this chapter is to assess the possible strategies of electrical stimulation of the colon by testing different electrodes, sites and configurations and by determining the optimal parameters of stimulation, such as amplitude, pulse width, frequency, duration of stimulation, and quantity of pulses. These values were first identified with *in vitro* experiments, described in the second section, and then tested with *in vivo* acute experiments on the pig, described in the third section. Similar induced contractions have been reproduced using an implantable device, which is a prerequisite for human application. Preliminary results of chronic experiments, always on the pig are described in the fourth section.

9.1 CTI Project Presentation

The ColoStim project proposes a new approach to the problem of motility disorders. Its objective is to develop an implantable electrical stimulator of the colon for the treatment of chronic constipation. ColoStim is a collaboration project between Medtronic Swiss R&D, the Divisions of Surgery, Gastroenterology and Anaesthesiology of the Lausanne University Hospital, the Signal Processing Institute of the Swiss Federal Institute of Technology of Lausanne, and the Department of Physiology of the University of Lausanne. This project was founded by the Swiss Government CTI MedTech initiative project number 4921.1 MTS and by Medtronic Europe. The project started on April 1st, 2001 with a grant for three years. It was mainly divided into two phases:

Phase 1: Acute experiments on twenty pigs. Exhaustive investigation of the strategies of electrical stimulation of the colon: test of different stimulation sites on the colon, test of amplitude and frequency, multi-site stimulation, gastric stimulation.

Phase 2: Chronic experiments on five pigs. Further development of an optimal stimulation protocol and clinical validation on awake animals. Assessment of the impact of electrodes and implantable device.

During this project, a patent has been filed on April 23st, 2003: US Patent Application Number 10/422,077, "Electrical stimulation of the colon to treat chronic constipation", A. Camps, N. Virag, M. Bertschi, and P. Kučera.

This study was made possible by grants from the Swiss Governmental Commission of Innovative Technologies (CTI) and Medtronic Europe. Electrodes for this study were designed and manufactured by Toine Camps, Jean Rutten and Ron van der Kruk from Medtronic Bakken Research Center in the Netherlands.

9.2 In Vitro Experiments

In vitro experiments were performed with the experimental setup presented in the previous chapter, but with contractions induced by electrostimulation. These experiences helped us to estimate optimal parameters for intestinal muscle stimulation, as well as to identify the threshold values inducing a minimal muscular contraction. These values were obtained by measuring the effect of stimulation on the porcine colonic strip tissue using different pulse parameters.

9.2.1 Description of the Equipment

Fragments of colonic muscles from pigs were placed in a bath containing an oxygenated physiological solution as explained in the previous chapter and illustrated in figure 8.1. Each colonic fragment was fixed to the transducer measuring the amplitude of mechanical strength. Two types of samples were used in this experiment: longitudinal and circular (transverse) strips. Electrical stimulation was applied using silver–chloride electrodes designed by the Physiology Department and inserted into the muscular layer fixed with physiological glue. Several experiments of electrical stimulation were performed in order to determine the optimal stimulation parameters leading to a contraction of the porcine colonic fragment.

A Grass S9 (Grass Medical Instruments, Kings Park, NY) device was used to generate electrostimulations. This external stimulator can generate monophasic and biphasic waves in the parameter range: 1–50 Hz, 1–100 ms pulse width, 1-30 V amplitude and duration up to 90 s.

9.2.2 Results

Several experiments of electrical stimulation have been performed in order to determine the optimal stimulation parameters, such as amplitude, pulse–width, frequency, and number of pulses, leading to a sufficiently strong tissue contraction. Figure 9.1 shows an example of *in vitro* electrical stimulation with the Grass S9 stimulator. It is interesting to note that the responses of the colonic tissue, illustrated on the left of figure 9.1, are dependent on the intensity and duration of stimulations. Pulse width and amplitude are thus critical parameters, while changes in frequency do not produce significant changes. We also observed that there is a refractory period: if an electrical stimulation is repeated too early, the intensity of the second contraction is lower. The interval between two successful stimulations are in the order of 1 min, which is close to the measured intrinsic frequency. According to our *in vitro* experiments, the optimal parameters are the following: monophasic stimulation, amplitude of 4 V, pulse width of 100 ms, frequency of 2 Hz, duration of pacing of 3 s.

A strength–duration curve has been created and is displayed on the right of figure 9.1. The three diameters of the disks represent the contraction amplitudes of the specified combinations of stimulation parameters. The biggest disks corresponds to normalized amplitudes in the range $(0.5, 1]$, the intermediate ones to the range $(0.25, 0.5]$, and the smallest one in $[0, 0.25]$. The

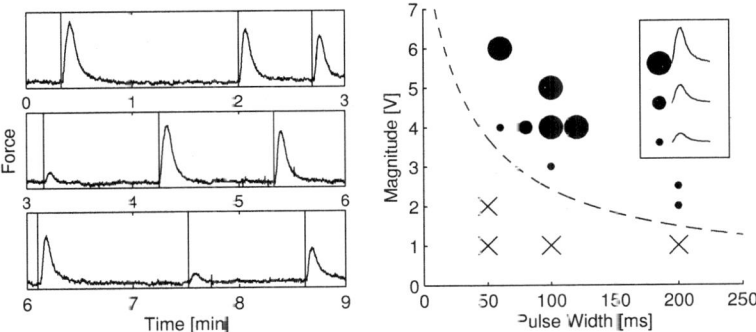

Figure 9.1: On the left: examples of contractions provoked by single pulse electrical stimulation (indicated with a vertical line). First line: the same monophasic stimulation 4 V, 100 ms is applied 3 times provoking similar responses. Second line: test of amplitude 3 V, 4 V and 5 V (monophasic pulses of 100 ms). Third line: test of duration 80 ms, 60 ms, 120 ms (monophasic pulses of 4 V). On the right: strength–duration curve for the electrical stimulation of a porcine caecum strip. Electrical stimulation with amplitude below this curve did not provoke a colonic contraction and is illustrated by ×, while the disks represent the successful stimulations, with size proportional to the amplitude of the response.

border is represented with a dashed line. Stimulations below this line do not provoke significant muscle contraction.

9.3 In Vivo Acute Experiments

Admittedly in specific conditions, the *in vitro* experiments have shown that the response of the intestinal tissue to an electrical stimulation is dependent on the intensity and the duration of the stimulus. Parameter values generating the strongest contractions, as well as their threshold values, have been identified [102]. The next step, and the objective of this section, is to evaluate the feasibility of intestinal electrostimulation *in vivo*. For this purpose, animals were anesthetized during all the experiment. Electrostimulations were applied to their intestinal wall and the muscular response was assessed in three different ways. The animals were eventually sacrificed. The pig had been chosen as the animal model because it is omnivorous and its weight and the size of its organs are comparable to those of a human. It should be noted that the anatomy of the pig's colon is different from the human's. However, even with all the numerous anatomical differences, such as the taenia, the transverse segment, the thickness of the tissue, etc., the pig remains the best animal model for that study.

9.3.1 Preparation of Animals

Experiments were performed on twenty healthy adult farm pigs (thirteen males and seven females) with an average weight of 44.9 ± 5.4 kg. All the animals were normally fed and had free access to water. They were operated under anaesthesia after overnight fasting.

Intramuscular Ketamine pre-medication (10 mg/kg) was used and general anaesthesia was induced by inhalation of Isoflurane via a muzzle mask. A short Teflon cannula was then inserted into the ear vein and connected to a perfusion of Alphachloralose (25 mg/ml). Induction of anaesthesia was terminated with a bolus dose of 100 mg/kg of Alphachloralose. A cuffed endotracheal tube, with an inner diameter of 9 mm, was then passed and the lungs were ventilated with a mixture of air and oxygen by a respirator. Oesophageal temperature was measured and maintained at 38 ± 0.5 °C by a heating element incorporated in the operating table. Anaesthesia was maintained with a continuous infusion of Alphachloralose at 30–40 mg/kg/h. Animals were constantly perfused with Hartmann solution at a rate of 3–5 ml/kg/h and their arterial blood pressure was monitored from a cannula inserted into the carotid. A lumbar epidural anaesthesia using a solution of 0–5% plain bupicvacaine (5–8 ml/h) was commenced before the median laparotomy, by which the colon, or at least a selected portion of it, was exposed. The exposed segment was regularly hydrated using a physiologic solution at 37 °C to avoid degradation of the colon state.

The type of general anaesthesia significantly influences the intrinsic colonic activity and the stimulation threshold. More specifically, Atropine and Dormicum had to be omitted since it has been shown in our *in vitro* experiments that they completely suppress intrinsic activity of the colon. The effect of Ketamine on the intestinal activity is also dramatic, as shown in the previous chapter. Amaris et al. similarly observed that electrically provoked ring–like contractions were abolished after the administration of Atropine [97]. Thus, special care was taken during anaesthesia in order to find the best way to make the pig asleep while keeping the enteric nervous system awake with a drug having minimal impact on intestinal transit. Three different anaesthetics, namely Isoflurane, Propofol and Alphachloralose were tested and best results in terms of colonic activity were achieved using Alphachloralose. This drug preserves the parasympathetic tone and allows the induction of colonic contractions.

Various specifically designed pacing–sensing electrodes were implanted by pairs in the subserosa at several locations and in different configurations. Electrodes were implanted using semi-circular needles, usually in the taenia region, and anchored in the subserosa with a non-resorbable string. Each pair could be connected to a standard implantable electrical stimulator device.

The magnet of the Magnet Tracking device was either fixed on the intestinal wall with a nonresorbable string, or introduced surgically in the lumen, permitting to determine if an electrical stimulation generated a movement or a contraction of the intestinal wall. A PVC prototype perfusion catheter was also surgically introduced into the lumen, allowing us to measure pressure variations provoked by electrostimulations.

At the end of the experiments, animals were sacrificed with an overdose of barbiturate.

9.3.2 Description of the Equipment

A description of the acute experimental setting is given in the following. The equipment to measure the effect of intestinal electrostimulation is first presented. Then, the two devices providing the electrical stimulations, as well as the different electrodes, are explained.

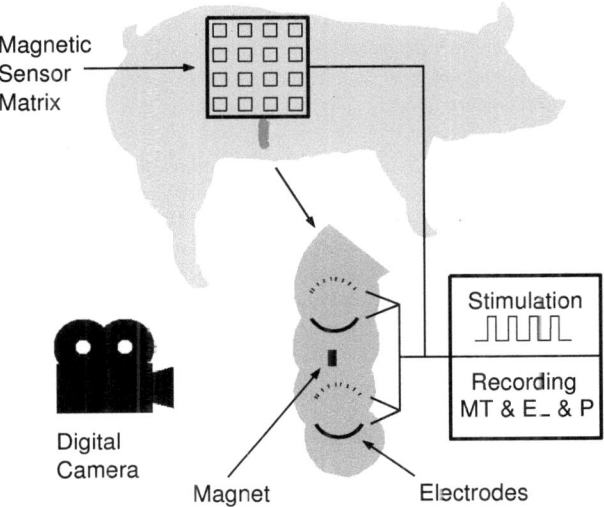

Figure 9.2: Description of the acute experimental setting to measure the effect of intestinal electrostimulations. The intestinal segment of interest, the caecum in this case, is exposed and recording–stimulation electrodes are fixed by pairs into its wall. These electrodes are connected to an amplifier, in order to measure the electrical activity (EL), or to an external implantable pulse generator, so as to generate electrostimulation. A magnet is inserted into the lumen, allowing us to assess the mechanical movements via an array of magnetic field sensors (MT) positioned on the abdomen. A camera is used to record visual contractions of the intestine. In some cases, a pressure catheter was also used to measure the intraluminal pressure (P). The signals collected from the colon are digitized and sent to a laptop for the off–line analysis.

Activity Assessment

Intestinal activity was assessed in four ways: visual inspection of contractions, evaluation of the contractions with the Magnet Tracking system, and measurements of the intraluminal pressure. Figure 9.2 describes the experimental settings for the acute experiments.

During experiments under complete anaesthesia, the intestinal segment was exposed and the visual effect of the electrical stimulations was recorded with a SONY digital camera. The camera was linked by IEEE 1394 (FireWire) to a personal computer, where periods of interest were selected and analysed. Stimulations were also subjectively evaluated by three persons with four discrete values: 0 corresponds to a nonperceptible contraction, 1 to a limited contraction, 2 to a moderate contraction, and 3 to a strong contraction. With this code, only the intensity is considered, without distinction of the type of the response, i.e longitudinal, circular, occlusive, propagated. In some cases, circular blue markers were glued on the intestinal wall, so as to automatically capture contractions with an off–line analysis. Recorded images were post–processed and tracking of the markers on the video sequence was performed.

The Magnet Tracking system was used to determine if an electrostimulation provoked movements or contractions of the intestinal wall. The basic principles of this novel device have been presented in the previous chapter.

In addition to the assessment of electrical activity and mechanical movements, the intraluminal pressure was also measured. For this purpose, a PVC prototype perfusion catheter (Sedia AG, Givisiez, Switzerland) was introduced in the intestinal lumen, allowing us to monitor the pressure variations provoked by the electrostimulation. The manometric probe (3.9 cm diameter) consists of four pressure transducers spaced 5 cm apart. Normal saline solution was perfused (0.5–1 mL/h) and the pressure was monitored in real time (SpaceLabs Inc. model number 90303, Redmond, Washington, USA).

Electrical Stimulation

Two different types of electrical stimulation devices were used for *in vivo* experiences. In the first stage, the Grass S9 (Grass Medical Instruments, Kings Park, NY) stimulator was used. It is the same external device as for *in vitro* stimulation tests. In the second stage, a battery–operated Medtronic 3625 neurological test stimulator device was used. This external device is employed in clinical practice to test the effectiveness of stimulation intraoperatively before implantation. It has the same features as an implantable stimulator. Other stimulation parameter values selectable with this stimulator were tested, so as to obtain contraction responses similar to those obtained with the Grass S9 stimulator. Maximum pulse width, in this case 1 ms, is a limit value for the implantable device. Monophasic square waves can be generated with a frequency range of 1–120 Hz and an amplitude range of 1–10 V.

Tested electrodes were of different lengths (2–40 mm) and composition. These electrodes need to fit the specific constraints related to the intestine. They have to be flexible enough to follow the anatomy of the tube and not restrict its movements. Stainless steel electrodes are widely used in myocardial pacing. However, it is known that low electrode impedance is the key factor for minimal consumption, which is why platinum–iridium electrodes were also tested.

The Medtronic 6500 unipolar temporary myocardial pacing leads are stainless steel 2 mm–long electrodes. The Medtronic 6494 unipolar temporary pacing leads are also stainless steel electrodes, but 40 mm–long. The Medtronic *ColoStim* are specially customized platinum–iridium electrodes 40 mm–long. Patch electrodes have also been tested, but they are difficult to fix and do not stay in contact with the intestinal wall during contractions.

During electrical stimulation, impedance and delivered current was also measured with an oscilloscope.

9.3.3 Description of the Data Processing Methods

The study of the GIS by imaging has a long history. Researchers, principally physiologists, developed several methods and techniques to observe and visualize the motility of the intestine using video sequences [103]–[104]–[105]. The algorithm used to automatically track coloured markers glued on the intestinal wall is presented next.

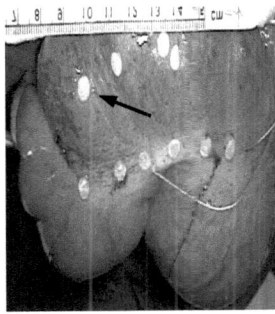 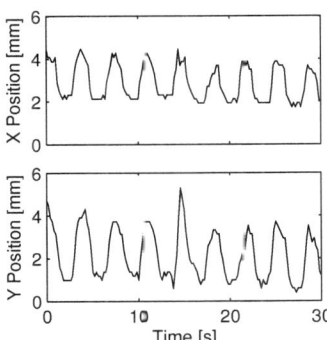

Figure 9.3: On the left, a frame of a video recording. Disk-shape markers are glued on the intestinal wall for the intestinal movement assessing. The X–Y position tracking of the marker indicated with the black arrow is displayed on the right. Oscillations at about 18 bpm correspond to respiration.

Several marker dots were glued on the serosal surface as illustrated on the left of figure 9.3, in order to quantify local contractions. A digital video camera was used to record the movements of the markers for periods of 5–10 min, and the video sequence was decomposed as a sequence of consecutive frames. Those frames were subsampled by five for further processing, which essentially consisted in tracking the movement of the markers frame by frame in a two–dimensional plane. The distances were calibrated by recording a ruler, marked off in millimeters, at the beginning of each recording session.

A software routine was developed to track the position of the markers in every frame. Tracking of the marker positions can not be performed with a trivial colour thresholding, because of the unfavourable illumination conditions. Moreover, the intestinal wall had to be hydrated, causing a disastrous light reflection and colour variation. In order to simplify the task, we used blue markers, considered to be sufficiently different from the colours of the others in the image. The marker to be tracked is manually selected in the first frame of the video sequence. The starting position of the marker P_0, as well as the RGB values of its colour C_0, are stacked in memory. The next position of the marker is evaluated by finding a neighborood region with a colour similar to C_0. For this purpose, the euclidian distance in the RGB space, between the colour of each pixel of the frame and C_0, is computed. Only the pixels at a distance less than a certain threshold are considered. This threshold is strictly related to the video sequence illumination. A morphological closing operator, the constructor element being a small disk, is performed on the thresholded frame, and filled regions of interest are obtained. Regions larger than a certain area are rejected, because of the saturation effects on some parts of the image due to the unfavourable illumination. The mass centre of each remaining region is then evaluated and the region closest to the previous marker position P_0 is considered as the new position P_1. If the colour C_1 is not updated from that of C_0, the marker colour variation, always due to the illumination, is not taken into account and the algorithm will quickly fail. The colour of the pixel at position P_1 may be not representative of the marker colour either, because of a shadow or colour saturation. The colour C_1 is thus evaluated as a weighted sum of C_0 (weight 0.8) and the the mean value of the colour of the region having P_1 as mass centre (weight 0.2). A maximal translation of the marker is fixed to avoid errors when the region corresponding to the tracked marker is not detected. In this case position and colour remain unchanged and the following frame is processed. Marker tracking is completed by repeating these steps for each frame of the video sequence. Several markers can be tracked at the same time.

An example of marker tracking on a porcine caecum is illustrated on the right of figure 9.3. The temporal evolution of the position P is decomposed into its two components x and y, that is plotted for a period of 30 s. The oscillations at about 18 bpm are due to the artificially induced respiration of the animal. Note that the good resolution of the frames give rise to an accurate spatial tracking, with a resolution of 0.2 mm.

9.3.4 Results

Several segments of the porcine colon have been electrostimulated, but the caecum was finally chosen as it proved to be easily accessible in the pig and had more similar characteristics to the corresponding human organ section.

Electrode pairs were fixed within the caecal wall, usually in the taenia region, that presents an adequate thickness for implantation. Electrodes of different compositions were tested and better results were achieved with platinum–iridium. Various electrode lengths were tested, as well as different configurations. Best results seem to have been obtained when 40 mm–long platinum–iridium electrodes, the Medtronic *ColoStim* electrodes, were placed diametrically, opposite to each other, on both sides of the intestinal tube, in a circumferential configuration. The spacing between the two electrodes had to be about 10–30 mm.

The stimulation parameters identified in the *in vitro* experiments had no relevance when applied *in vivo*. Due to the drugs used for the anaesthesia, higher potential differences (up to 20 V) had to be used to provoke a muscular contraction of the caecum. The Grass S9 external stimulator was used to deliver these stimuli.

The effect of the electrostimulation of the caecum was first assessed visually. An example of provoked contraction is shown in figure 9.4. Propagation always follows a similar pattern: a strong occlusive contraction is generated when an electrical stimulation is applied. It starts to propagate about 30–60 s after the beginning of the stimulation. The largest propagation wave occurred during the relaxation phase of the colon, when electrical stimulation was stopped, lasting between 1–2 min and propagating approximately for 5 cm. A similar phenomenon was observed by Moritz et al. [106]. *In vivo* dog experiments showed on ten tests that the pressure continues to increase after termination of the electrical stimulus, a phenomenon called off–response. Some years later, Hughes et al. [94] also observed the same behaviour. The dog intestinal response has a latency of about 25 s.

When stimulation was applied with the Medtronic 3625 battery–operated device, amplitude and pulse width proved to be the critical parameters. Other stimulation parameter values, compatible with implantable devices, were identified by testing several combination of amplitude, frequency, and pulse width. Similar contractions were obtained using 8–10 V, 120 Hz, and 0.2–1 ms. Therefore, the main conclusion is that an implantable device can deliver sufficiently strong stimuli eliciting contractions in the caecum.

A strength–duration curve was established for the *in vitro* experiments. It is displayed in

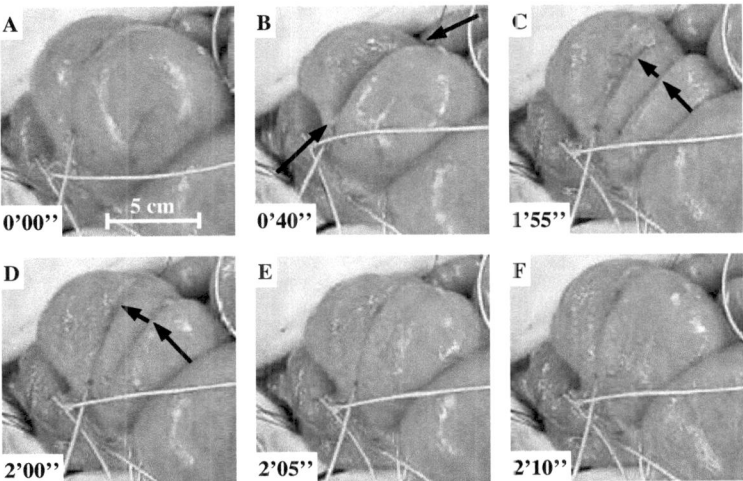

Figure 9.4: Example of a provoked 5 cm–propagated contraction in the caecum. (A) Initial state of the exposed caecal segment. (B) A pair of customized electrodes, Medtronic *ColoStim*, is implanted on both sides of the caecum. The stimulation parameters used for this experiment are: 20 V, 2 Hz, 100 ms during 1 min. The two black arrows indicate the position of the provoked local contraction. (C)–(D)–(E) About 1 min after the electrostimulation, a local contractions is generated and propagate approximatively for 5 cm. (F) The caecum relaxes and come back to the initial situation.

figure 9.5. A careful combination of parameters (amplitude and pulse width) was used for each applied electrostimulation, and its effect on the caecum was visually evaluated by four persons, with the evaluation score described above. These scores are associated to four graphical symbols, i.e. × and three disks with different diameters, and are represented in figure 9.5. The larger the disk diameter the stronger the response. The estimated borders between successful and unsuccessful stimulation cases is represented with a dashed line. Electrostimulation of the porcine caecum below this limit did not provoke muscle contraction.

The amplitude threshold to generate an occlusive contraction is 8 V for a pulse width of 1 ms, and the minimum pulse width to generate a contraction is 200 μs for an amplitude of 10 V. Changes in pacing frequency do not induce significant differences, but higher frequencies tend to provoke stronger contractions. The optimal parameter set is 10 V, 1 ms, 120 Hz. The injected current must be in the range of 7–15 mA. It has to be noted, that the stimulation parameters we considered as optimal correspond to values at the limit of deliverability for the implantable device. Stimulations must be applied for 5 to 90 s, with longer stimulations provoking stronger contractions, and the apex of the contraction is reached after about 30–60 s.

It is important to note that these parameter values are of the order of magnitude of the published parameters presented in table 9.1. Moreover, Moritz et al. [106] established that the dog tolerates the current necessary to evoke contractions in the range of natural contraction. Further

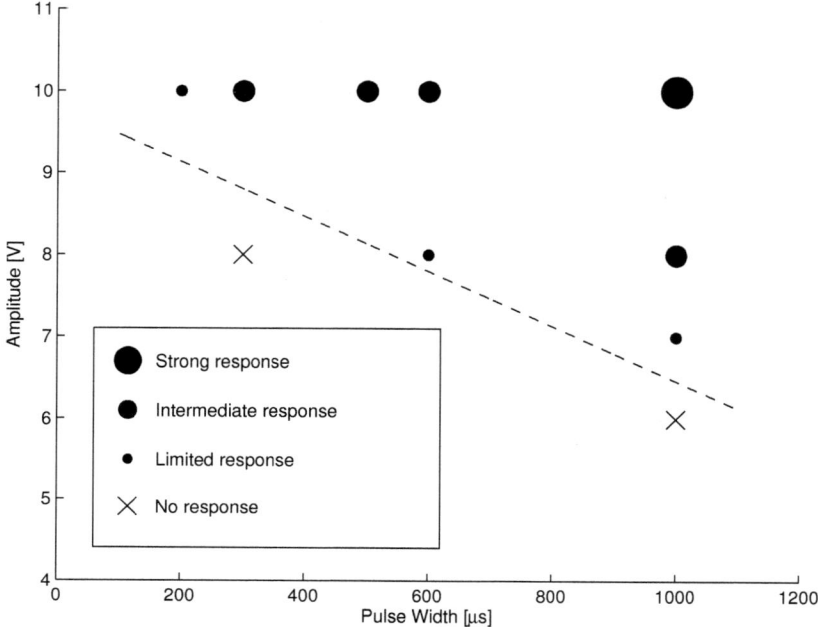

Figure 9.5: The dashed line indicates the strength–duration curves for the 120 Hz electrical stimulation of the porcine caecum. *In vivo* experiments are effectuated under Alphachloralose anaesthesia and using a battery–operated device. Electrical stimulations with amplitude below this line did not provoke a colonic contraction (marked by ×). The point size is proportional to the visual response of the electrostimulation.

studies accomplished by Accarino et al. [107]–[108], showed that in twelve healthy humans, electrostimulation is perceived when the intensity is about 39 mA (15 Hz and 0.1 ms), and becomes uncomfortable at about 63 mA. Currents delivered by our battery–operated device during stimulations are considerably smaller than both thresholds.

Visual inspection provides interesting results, but it is a subjective process and, from an objectivity viewpoint, not entirely convincing. In order to quantify local induced contractions, we use the technique of markers tracking presented above. Figures 9.6 shows an example of marker tracking during a caecal induced contraction. Four frames of the video sequence are displayed. The euclidian distance between the pixel colours of each frame and that of one tracked marker are represented. Marker extraction was done by a simple thresholding of these grayscale pre–processed images. The position of two markers, represented in the four frames as black spots, are detected and the distance r between them is computed and plotted with respect to time. The advantage of considering the distance between the two markers and not their absolute position, is that external perturbations, e.g. respiration, are directly subtracted and only the

Section 9.3 — In Vivo Acute Experiments

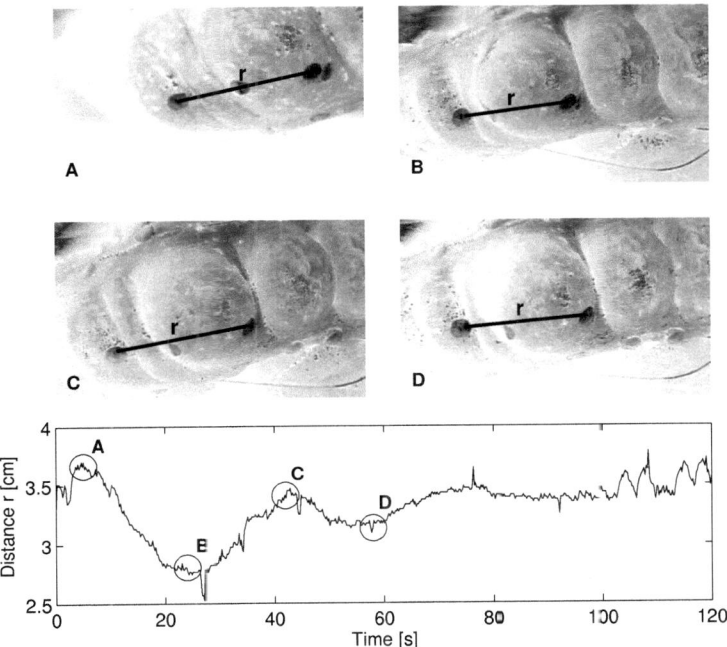

Figure 9.6: Four frames are extracted from the video sequence of an induced caecal contraction (A-D) and preprocessed, so as to identify the markers. Positions of two black dots (markers) are computed, and the distance r between them is plotted with respect to time. Circles on the curve indicate the instant corresponding to frames ABCD.

local effects of the stimulation are appraised.

The induced contraction translates into an oscillation. The electrical stimulation starts at position A and 25–30 s later. in correspondence with the end of the stimulation, the first and maximal local contraction take place at position B. The two markers get closer from about 3.7 to 2.7 cm. Markers are placed in the longitudinal direction and thus the variations of the distance r correspond to a longitudinal contraction of the muscle. A decrease of r up to 1 cm over the initial 3.7 cm corresponds to a longitudinal muscle contraction of about 30% of its stretched length. From position C to D a second local contraction wave is generated, weaker than the previous one but easily perceptible. About 3 minutes after the beginning of the stimulation, the initial marker distance r is recovered.

Visual inspection is a direct evaluation of stimulation efficacy during acute experiments, but is not applicable for *in vivo* chronic experiments because of its invasiveness. The Magnet Tracking system allowed us to observe non–invasively the gastro–intestinal motility by the tracking of an ingested magnet. We used this novel technology to detect if the electrical stimulation induced a

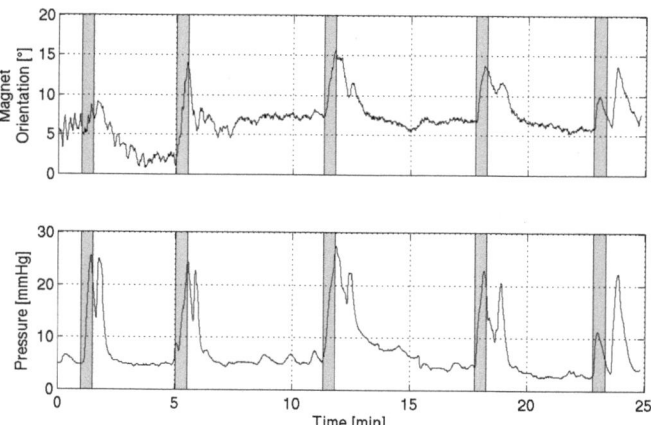

Figure 9.7: Simultaneous mechanical recording with Magnet Tracking (top) and intraluminal pressure (bottom) during electrical stimulation. Each vertical gray bar represents the pacing with appropriate parameter values: 10 V, 1 ms, 120 Hz during 30 s followed by a relaxation phase of about 4–5 min.

movement of the caecal wall. The magnet was positioned in the region of electrode implantation and its displacement, induced by the electrical stimulation, was detected by the device. We simultaneously measured the intraluminal pressure changes at the site of the magnet movement. The signals recorded with these two complementary methods are presented in figure 9.7 for five electrical stimulations. Five stimulations were applied and five local contractions were induced. Note that both modalities always record a two–peaks waveform, starting with the beginning of the stimulation, reaching a first peak at the end of the stimulation. A second one was observed about one minute later, with a final return to the resting state 2–3 minutes after the beginning of the movement. Some one–peak waveforms were also observed, but no evidence on their cause was found.

An increase of intraluminal pressure of 20 mmHg was measured, corresponding to the values reported in the literature for an intrinsic contraction of the intestine [109]. Moritz et al. [106] observed an increase of 10–85 mmHg in the dog small intestine after electrostimulation of the tissue. Another interesting but negative feature found out is that contraction amplitude decreased gradually in time. Also a tendency to degradation of the intestinal tissue could be observed visually.

The measure of the intraluminal pressure by a special transducer is the only way to validate the efficacy of electrical stimulation and to compare the results with those of the literature. Manometry is an invasive technique and recordings are very uncomfortable. On the other hand, the Magnet Tracking is totally non–invasive, but it does not give information about the amplitude of the contraction. The simultaneous pressure–Magnet Tracking recordings showed that induced

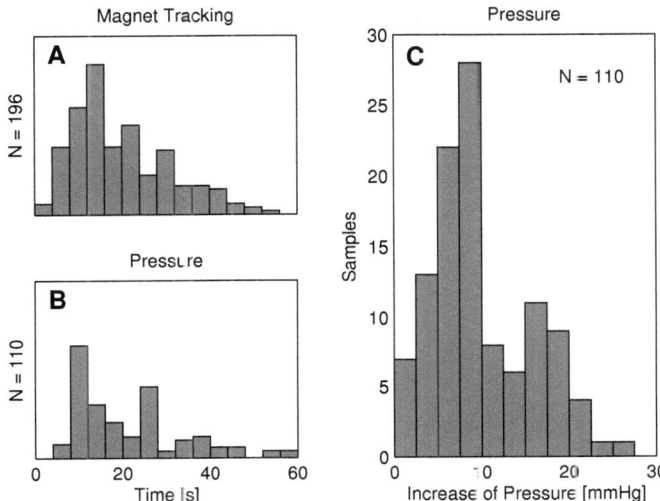

Figure 9.8: Statistics corresponding to 4 pig experiments. The time interval between stimulation and maxima is represented in A for Magnet Tracking and in B for pressure measurements. Distribution of the increase of pressure between the baseline pressure measured before the stimulation and the maximum value reached during or after the latter is presented in C.

local contractions are easily detected by both techniques. Thus, if the contraction amplitude estimate is required manometry has to be used, but if only detection is considered, the Magnet Tracking system is preferable.

The signals recorded during the stimulation sessions of four pigs were collected and a statistical analysis were performed. The efficient parameter values 10 V, 1 ms, 120 Hz during 30 s were imposed and the time interval between the stimulation beginning and the first signal maxima is observed. Figure 9.8A shows an histogram of the Magnet Tracking measures for 196 equal stimulations. Most of the induced responses occur within the 40 s that follow the beginning of the stimulation, with a peak between 15–20 s. The same analysis has been done with the manometric measurements, (figure 9.8B). The time interval between the stimulation onset and the first pressure maximum is evaluated for 110 stimulations. The observations are grouped in the first 30 s, but this time the histogram exhibits two modes, the first one around 10 s and the second one at about 25 s. The presence of this second mode may be due to the fact that the first peak of the induced contraction has not been detected, but only the second one. Figure 9.8C illustrates the histogram of maximal pressure increase measured after a contraction. Most stimulations induce a pressure increase of about 10 mmHg, but a group of stimulations reaches 20 mmHg, and in rare occasions, up to 30 mmHg.

9.4 *In Vivo* Chronic Experiments

The *in vivo* acute experiments have shown the feasibility of the electrostimulation of the porcine caecum using a battery-operated device. Parameter values compatible with implantable device limitations have been identified, and proved to induce intestinal local contractions. The next step of this project consisted in the investigation of the following three points: test of the sensibility of an awake animal to electric stimuli, possible electrode rejection, and verification of the efficacy of the electrostimulation. In this section, some preliminary results on the *in vivo* chronic experiments are presented.

9.4.1 Preparation of Animals

Experiments were performed on five healthy adult farm pigs (one male and four females) with an average weight of 42.4 ± 6.4 kg. All the animals were normally fed and had free access to water.

The chronic surgical procedure was performed in fully sterile conditions. Pigs were operated using the same anaesthesia protocol as for the acute experiments after overnight fasting. Following a median laparotomy, three pairs of electrodes were implanted in the caecum (diametrically opposite) with a spacing of 5 cm between electrode pairs. Electrodes were then guided to the back of the pig's neck. To be able to assess colonic movement on an awake pig, a magnet attached to a string was introduced by ileotomy and through the ilecocaecal valve inside the caecum, and sewed to the colonic wall. Before closing the abdomen, stimulation tests were performed for each pair of electrodes to check the response of the colon to electrical stimulation, so as to control their operativeness. Animals were transferred to a recovery cage at the farm and were coated with a specially designed jacket to protect the open wound on the neck and the electrode connectors. They were normally fed, while injection of antibiotics was continued until the sacrifice of the animals.

Pigs for chronic experimentation were trained to be placed in a hammock some days before the operation. The first chronic experiment was done under the supervision of the ethical committee to assess if pain was generated by electrical stimulation.

A week after electrode implantation, animals were sacrificed using barbiturate overdose, and autopsied.

9.4.2 Description of the Equipment

The experiments were performed with the experimental setup presented in the previous chapters. The Magnet Tracking system was used to detect the movements of the magnet induced by electrostimulation provided by the Medtronic 3625 device. An oscilloscope was used to measure the delivered current and also to evaluate if the electrodes were still located in the serosa.

9.4.3 Results

During the chronic experiments, we faced several technical and clinical problems. The intestinal tube is continuously moving in the abdomen, possibly causing entanglements of the tube and cables, or a tension on the electrodes. The first mishappen fortunately never took place. In fact, the autopsy showed that some days after electrode implantation the electrodes seemed glued on

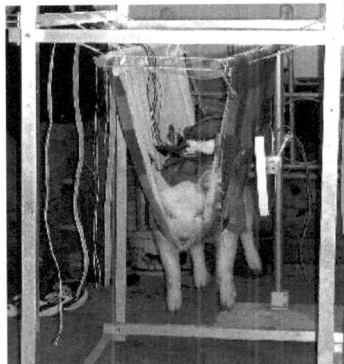 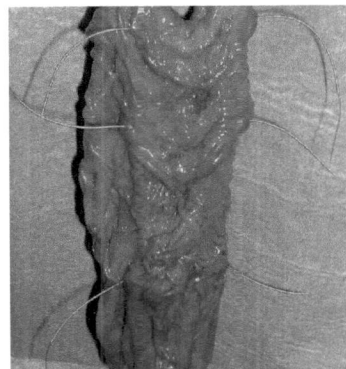

Figure 9.9: On the left, awake pig in hammock during a stimulation session. The Magnet Tracking sensor-array is positioned vertically near the animal, at the right of the image. On the right, view of the inside of the porcine caecum during the autopsy of a chronic experiment. The six Medtronic *ColoStim* electrodes have migrated in the caecal lumen.

the intestine. They were covered by fibrin patches about 10 cm long, as they were englobed in the intestinal wall. We observed that the tension on the cables generated by intestinal movements could cause a 1–2 cm delocalization of the electrodes. This could be a problem if the electrodes were in contact and caused a short-circuit during stimulation, but this could be simply solved by allowing a sufficient spacing between electrode pairs.

Special care had to be taken during surgery and recovery period because of the infection problems. Infection was observed in one experiment, probably due to the open wound on the neck or by a colonic perforation caused by the electrodes. This problem was solved for all subsequent animals by the use of antibiotics.

After implantation of the electrodes, the animal was covered with a special protective jacket, as shown in figure 9.9, and transferred to the farm. The use of the jacket had various objectives. It protects the wound and the electrodes from dirt, prevents the animal from tearing off the electrodes, permits an easy access to the electrodes and facilitates medical manipulations.

In general, the animal stayed quiet in the hammock during stimulation and measurements, as can be seen on the left of figure 9.9. However, many experimental sessions were perturbed because the animal was restless during the first day in the hammock. Getting the pig used to the hammock for some days before electrode implantation solved this problem.

Electrical stimulation experiments were performed at the farm with the animal in the hammock and no premedication. They were stimulated 2–3 days after electrode implantation, when they seemed to have recovered from the operation. Amplitudes up to 10 V were delivered and the conscious animal did not seem to feel the stimulation.

To avoid the detachment of the magnet, as observed in two experiments where the animal expelled it three days after the implantation, it had to be fixed to the intestine with a sufficiently thick and short string. Better signals were measured with animals of reduced weight and using a bigger permanent magnet.

For all the experiments, the implanted Medtronic *ColoStim* electrodes did not respond to stimulation 3–5 days after implantation. The autopsy revealed that all 6 electrodes had migrated inside the caecum lumen, as can be seen on the right of figure 9.9. Since some migration problems have been observed with the customized electrodes in the chronic experiment phase, electrodes Medtronic 6500 had to be tested again to study if they also present migration problems.

9.5 Conclusion

The activation of the intestinal motility for the treatment of slow transit constipation by electro-stimulation is a challenging goal. Our study shows that it is possible to provoke local contractions of the caecum with a pair of electrodes fixed on the serosa and connected to a battery–operated stimulation device.

In vitro experiments helped us to determine the ranges of stimulation parameters leading to a mechanical response of the intestinal tissue.

Published studies on the electrical stimulations of the small intestine and the colon show a wide dispersion of stimulation parameter values. *In vivo* acute experiments showed us that local caecal contractions can be provoked using a battery–operated device, the Medtronic 3625. One typical stimulation consisted of 30–60 s pacing at 10 V amplitude (corresponding to 15–20 mA), 1 ms pulse width and 120 Hz. The anesthetic strategy was designed so as to avoid factors decreasing intestinal motility.

Our study provides detailed information on the use of three different methods to assess the intestinal motility. Visual inspection is a direct way to assess colonic contraction, but image sequences are often difficult to process and not always available in acute experiments, and impossible in chronic experiments. An image processing method to track markers glued on the intestinal wall was presented, and results show the particular two–peaks waveform of the electrical induced contraction. Intraluminal pressure measurements are important for the comparison with other researches in the field, but provide limited informations. Intraluminal pressure variations are linked to signals recorded with the Magnet Tracking system, which allows us to observe the displacement of a permanent magnet in the gastro–intestinal tube.

The specially designed Medtronic *ColoStim* platinum–iridium electrodes were used to ensure a low impedance and allow maximal energy transmission to the muscle. *In vivo* chronic experiments revealed that these electrodes migrate inside the caecum lumen after 3–5 days after their implantation. The preliminary results on chronic experiments show that our stimulation parameters are below the threshold of perception and discomfort of the animal.

Chapter 10

Conclusion

The purpose of this research work has been the development of methods to assess intestinal activity and the construction of a model reproducing important features of the electrical and mechanical dynamics, along with the investigation of the electrostimulation of the colon provoking muscular contractions. Certain questions regarding the functions of the gastrointestinal system have been completely or partially answered, but some still remain open. The key points of this work are now summarized and followed by a review of possible future extensions.

Summary of Achievements

The introductory chapter has highlighted the impact of slow transit constipation, considered as one of the major symptoms among digestive disorders. The limits of conventional and experimental approaches have been identified. The emergence of computer simulations in this field of research, with their advantages but also their practical difficulties and limits, has been presented.

A brief overview of the major elements of the gastro–intestinal smooth–muscle cell was presented in chapter 2. Over this chapter, the terms, principles, characteristics, properties, and the processes related to the gastrointestinal smooth–muscle tissue were introduced. In particular, the cellular membrane and its electrophysiological specificity was detailed, in order to introduce the different processes involved in the substance transport across the membrane. The methods for measuring the ionic channel currents generated by the passive and active transports were also presented, as well as the description of the monopolar and bipolar techniques for measuring the membrane potentials and their propagation along the membrane. The particularities of the electrical activity of the gastrointestinal smooth–muscle cell have been treated, and the slow waves and spike potentials have been described. Some basic concepts described in this chapter were used in chapter 3 from a more global point of view. The elementary aspects of the structure, the functions, and the control of the gastro–intestinal system were briefly presented. The structure of the intestinal wall, the different sections of the digestive tube, and its associated accessory organs have been described along the path of the ingested food, from mouth to anus. Motility, secretion, digestion, absorption, and immuno–defense have been described as digestive functions, and their role in the complex digestive process has been presented. We also described the complex regulation of the gastro–intestinal system by combination of nervous, mechanical, and hormonal signals.

The mathematical basis of electrophysiological modeling has been introduced in chapter 4. We implemented three computer models to simulate the intrinsic electrical activity of the intestine. From the computational point of view, the Bardakjian model has the big advantage to be simple. It represents the human colon as a set of coupled oscillators on a tubular structure. The values of the intrinsic oscillator parameters are chosen in such a way to reproduce the observed *in vivo* features of the colon. If too simple, a model may fail to capture the salient features of the real system and have limited predictive ability. The recent progress on experimental techniques and computational power allowed us to employ realistic and complex models, such as the Miftakhov cellular model based on a description of transmembrane ionic currents. This complex mathematical model of the myoelectrical activity of the small intestine functional unit is based on real morphological and electrophysiological data. It provides the most elaborate and realistic way to model the slow waves and the spike bursts of the intestinal smooth muscle. This model gives interesting results in pharmacological experiments with single functional units, but is computationally too heavy for larger scale simulations. The computational cost of adding such a huge amount of details indeed inhibits the use of such a model. It is therefore usually appropriate to make some approximations to reduce the model to a reasonable size, as is the case for the model proposed by Aliev. This cellular model of the intestine is constituted by two coupled layers of longitudinal muscle and interstitial Cajal cells. The dynamics of each layer is described using the FitzHugh–Nagumo formalism. Indeed, computer–models of the intestine have to evolve before becoming realistic enough, as is the case for heart tissue models.

Simulation of physical and natural phenomena usually implies a discrete integration of partial differential equations expressing a variety of physical values in time and/or space. The importance of temporal and spatial discretization has been discussed in chapter 5. Forward, backward, and mixed Euler schemes for the time integration were presented and their computational cost, stability, and convergence was discussed. The centered second order finite difference has been used to discretize the one–dimensional diffusion equation, because of its simplicity and efficiency. It has been shown that the discrete handling of time and space always introduces numerical instabilities and errors, propagating through time and being strongly amplified by the non–linear system dynamics. As an effect, we typically observed a reduction of the conduction velocities of slow waves in the tissue.

The numerical models provide a large amount of data that have to be processed in order to permit a comparison with published experimental or clinical observations. Chapter 6 detailed some techniques for the processing of simulated results so to confront them to real measurements. It was shown that modeling the electrical activity by a set of coupled oscillators provided a qualitative agreement with the experimental observations. Space–time plots have been widely used to analyse experimental measurements, and they constitute also a useful and appropriate representation to visualize the behaviour of the oscillatory systems, such as the intestinal model. The spatial organization of the different segments of intestine and its time evolution is a basic but challenging question for researchers. That is why we proposed a simple and robust tool, based on the cross–covariance analysis, for the temporal tracking of oscillator dynamics. This new graphical tool has been applied to the Aliev computer–model. Moreover, electrograms of the intestinal tissue were modeled and discussed. The mass–displacement of a point-shape object was also modeled using the simulated intestinal electrical activity and assuming that slow waves are always accompanied with spikes, and thus always associated with a muscular contraction. A simple 3D reconstruction of the colon was presented, so as to permit a visualization of the simulations.

The electrical activity of the gastro–intestinal tract was investigated from the experimental viewpoints and the computer model in chapter 7. Differential recordings were performed *in vivo* by implanting electrodes in the serosa of three different segments of the porcine tube: the stomach, the caecum, and the sigmoid. In parallel, the simulated electrical activity using three versions of the Aliev computer-model was presented. Space-time plots were used to visualize slow-wave propagation and pacemaker location evolution in these models. The self-organization properties of the models were studied using the cluster evolution analysis. Electrograms were also computed using a dipole model, showing a strong similarity with experimental measurements.

The assessment of the mechanical activity of the gastro-intestinal tract was performed with *in vitro* and *in vivo* experiments, and results were presented in chapter 8. *In vitro* experiments allowed us to observe the spontaneous circular and longitudinal smooth muscle activity of a colonic porcine strip, and evaluate the effect of physiologically relevant agents on the mechanical activity of the tissue. *In vivo* measurements were performed with the Magnet Tracking, permitting to track an ingested magnet from the mouth to the anus. A powerful signal processing technique based on morphological filtering for the extraction of instantaneous frequency of the rhythmic contractions from Magnet Tracking signals was also presented. Synchronous recordings of the electrical and mechanical intrinsic activity of the porcine sigmoid in *in vivo* acute experiments have also been performed, and the effect of Ketamine on intestinal motility was observed. In parallel to animal experimental work, we used the computer model of Aliev to reproduce the experimental signals obtained with the Magnet Tracking system. Intrinsic electrical activity and respective mass displacement simulation have been visualized on a 3D geometrical model of the colon.

In chapter 9, the possibility to provoke local contractions of the caecum with a pair of electrodes fixed on the serosa and connected to a battery-operated stimulation device was demonstrated. *In vitro* and *in vivo* (acute and chronic) experiments were performed on the porcine gastro-intestinal tract, and typical stimulation parameters were identified, consisting of pacing during 30–60 s at 10 V amplitude (corresponding to 15–20 mA), 1 ms pulse width and 120 Hz. Detailed information on the use of three different methods to assess the intestinal motility was provided: visual inspection and a related image processing approach, intraluminal pressure measurements, and the tracking of a permanent magnet in the gastrointestinal tube (Magnet Tracking).

Future Work

The following non-exhaustive list contains of propositions for further extension of the work presented in this thesis:

- Study of slow waves and spike propagation in 2D tissue patches using the more realistic Miftakhov model.

- Further model validation, which is a critical and imperative point.

- Development of a more realistic geometrical model of the human colon for the visualization of simulations. This could give more information about the intestinal activity.

- Development of a virtual electrostimulation in the model, in order to test the experimental parameter values and investigate new stimulation protocols.

- Development of innovative processing approaches for Magnet Tracking signals, based for instance on blind source separation.
- Finally, electrostimulation of the colon has to be improved with chronic experiments, particularly by testing the electrode migrations and long–term stimulation efficacy.

Bibliography

[1] A. C. Guyton and J. E. Hall. Contraction and excitation of smooth muscle. In *Textbook of medical physiology*, chapter 8, pages 95–103. W. B. Saunders Company, 1996.

[2] A. C. Guyton and J. E. Hall. Transport of ions and molecules through the cell membrane. In *Textbook of medical physiology*, chapter 4, pages 43–55. W. B. Saunders Company, 1996.

[3] A. C. Guyton and J. E. Hall. Membrane potentials and action potentials. In *Textbook of medical physiology*, chapter 5, pages 57–72. W. B. Saunders Company, 1996.

[4] B. L. Bardakjian. Gastrointestinal system. In J. D. Bronzino, editor, *The biomedical engineering handbook*.

[5] A. C. Guyton and J. E. Hall. Transport and mixing of food in the alimentary tract. In *Textbook of medical physiology*, chapter 63, pages 803–813. W. B. Saunders Company, 1996.

[6] A. C. Guyton and J. E. Hall. General principles of gastrointestinal function — motility, nervous control, and blood circulation. In *Textbook of medical physiology*, chapter 62, pages 793–802. W. B. Saunders Company, 1996.

[7] A. C. Guyton and J. E. Hall. Secretory functions of the alimentary tract. In *Textbook of medical physiology*, chapter 64, pages 815–832. W. B. Saunders Company, 1996.

[8] A. C. Guyton and J. E. Hall. Digestion and absorption in the gastrointestinal tract. In *Textbook of medical physiology*, chapter 65, pages 833–844. W. B. Saunders Company, 1996.

[9] W. A. A. Kunze and J. B. Furness. The enteric nervous system and regulation of intestinal motility. *Annu. Rev. Physiol.*, (61):117–142, 1999.

[10] B. L. Bardakjian and S. K. Sarna. Using computer models to understand the roles of tissue structure and membrane dynamics in arrhythmogenesis. *Proc. IEEE*, 84(3):334–354, March 1996.

[11] B. L. Bardakjian. *Analysis and modeling of the electrical control activity in human colon*. PhD thesis, McMaster University, Sept. 1978.

[12] B. L. Bardakjian and S. K. Sarna. A computer model of human colonic electrical control activity (eca). *IEEE Trans. on Biomed. Eng.*, BME–27(4):193–202, April 1980.

[13] S. K. Sarna, B. L. Bardakjian, W. E. Waterfall, and J. F. Lind. Human colonic electrical control activity (eca). *Gastroenterol.*, (78):1526–1536, 1980.

[14] D. A. Linkens, I. Taylor, and H. L. Duthie. Mathematical modeling of the colorectal myoelectrical activity in humans. *IEEE Trans. on Biomed. Eng.*, BME–23(2):101–110, March 1976.

[15] L. O. Chua and D. N. Green. Synthesis of nonlinear periodic systems. *IEEE Trans. on Circ. and Sys.*, CAS–21(2):286–294, March 1974.

[16] A. L. Hodgkin and A. F. Huxley. A quantitative description of membrane currents and its application to conduction and excitation in nerve. *J. Physiol.*, (117):500–544, 1952.

[17] D. Noble and R. L. Winslow. Reconstructing the heart: network models of sa node–atrial interaction. In A. V. Panfilov and A. V. Holden, editors, *Computational biology of the heart*, chapter 2, pages 49–64. John Wiley & Sons, 1997.

[18] D. A. Linkens and S. Datardina. Frequency entrainment of coupled hodgkin–huxley–type oscillators for modeling gastro–intestinal electrical activity. *IEEE Trans. on Biomed. Eng.*, BME–24(4):362–365, July 1977.

[19] D. A. Linkens. Electronic modeling of slow–waves and spike–activity in intestinal tissue. *IEEE Trans. on Biomed. Eng.*, BME–27(7):351–357, July 1980.

[20] M. R. Boyett, A. Clough, J. Dekansky, and A. V. Holden. Modelling cardiac excitation and excitability. In A. V. Panfilov and A. V. Holden, editors, *Computational biology of the heart*, chapter 1, pages 1–47. John Wiley & Sons, 1997.

[21] M. E. Belik, T. P. Usyk, and A. D. McCulloch. Computational methods for cardiac electrophysiology. In P. G. Ciarlet, editor, *Handbook of numerical analysis*, volume 12, chapter 2. Elsevier, 2004.

[22] R. N. Miftakhov, G. R. Abusheva, and D. L. Wingate. Model predictions of myoelectrical activity of the small bowel. *Biol. Cybern.*, (74):167–179, 1996.

[23] Z. S. Wang, M. Abo, and J. D. Z. Chen. Rhythms and chaos in the stomach. In M. Akay, editor, *Nonlinear biomedical signal processing*, volume 2, chapter 13, pages 319–338. IEEE Press, 2001.

[24] R. N. Miftakhov and G. R. Abusheva. Effects of selective $mboxK^+$-channel agonists and antagonists on myoelectrical activity of a locus of the small bowel. *Biol. Cybern.*, (75):331–338, 1996.

[25] R. N. Miftakhov, G. R. Abusheva, and J. Christensen. Numerical simulation of motility patterns of the small bowel. 2. comparative pharmacological validation of a mathematical model. *J. theor. Biol.*, (200):261–290, 1999.

[26] A. A. Hagberg. *Fronts and patterns in reaction–diffusion equations*. PhD thesis, University of Arizona, 1994.

[27] R. R. Aliev, W. Richards, and J. P. Wikswo. A simple nonlinear model of electrical activity in the intestine. *J. theor. Biol.*, 204:21–28, 2000.

[28] R. A. FitzHugh. Impulses and physiological states in theoretical models of nerve membrane. *J. Biophys.*, (1):445–466, 1961.

[29] R. FitzHugh. Mathematical models of excitation and propagation in nerve. In H. P. Schwan, editor, *Biological Engineering*.

[30] C. F. Gerald and P. O. Wheatley. *Applied numerical analysis*. Addison–Wesley Publishing Company, fourth edition, 1989.

[31] O. Blanc. *A computer model of human atrial arrhythmia*. PhD thesis, Lausanne Swiss Federal Institute of Technology, 2002.

[32] M. Kunt. *Traitement numérique des signaux*. Presses polytechniques et universitaires romandes, 1984.

[33] A. V. Oppenheim, R. W. Schafer, and J. R. Buck. *Discrete-time signal processing*. Prentice–Hall, second edition, 1999.

[34] V. Jacquemet. *A biophysical model of atrial fibrillation and electrograms: formulation, validation and applications*. PhD thesis, Lausanne Swiss Federal Institute of Technology, 2004.

[35] W. J. E. P. Lammers, E. Stephen, J. R. Slack, and S. Dhanasekaran. Anisotropic propagation in the small intestine. *Neurogastrointerol. Mot.*, 14:357–364, 2002.

[36] G. W. Hennig, M. Costa, B. N. Chen, and S. J. H. Brookes. Quantitative analysis of peristalsis in the guinea pig small intestine using spatio-temporal maps. *J. Physiol.*, 517(2):575–590, 1999.

[37] P. Berčík, L. Bouley, F. Dutoit, A. L. Blum, and P. Kučera. Quantitative analysis of intestinal motor patterns: Spatiotemporal organization of nonneural pacemaker sites in the rat ileum. *Gastroenterology*, 119(2):386–394, Aug. 2000.

[38] D. J. DeShazer, R. Breban, E. Ott, and R. Roy. Detecting phase synchronization in a chaotic laser array. *Phys. Rev. Lett.*, 87(4):1–4, 2001.

[39] S. Jalan and R. E. Amritkar. Self-organized and driven phase synchronization in coupled maps. *Phys. Rev. Lett.*, 90(1):1–4, 2003.

[40] J. Bhattacharya and H. Petsche. Enhanced phase synchrony in the electroencephalograph γ band for musicians while listening to music. *Phys. Rev. E*, 64(012902):1–4, 2001.

[41] S. Rzeczinski, N. B. Janson, A. G. Balanov, and P. V. E. McClintock. Regions of cardiorespiratory synchronization in humans under paced respiration. *Phys. Rev. E*, 66(051909):1–9, 2002.

[42] R. Q. Quiroga, T. Kreuz, and P. Grassberger. Event synchronization: a simple and fast method to measure synchronicity and time delay patterns. *Phys. Rev. E*, 66(041904):1–9, 2002.

[43] M. Bertschi, N. Virag, V. Schlageter, P. Kucera, J.-C. Givel, and J.-M Vesin. Analysis of cluster evolution in a model of intestine electrical activity. *Proc. Engineering in Medicine and Biology Society*, 3:2683–2686, September 2003.

[44] G. W. Botteron and J. M. Smith. A technique for measurement of the extent of spatial organization of atrial activation during atrial fibrillation in the intact human heart. *IEEE Trans. on Biomed. Eng.*, 42(6):579–586, June 1995.

[45] R. Q. Quiroga, A. Kraskov, T. Kreuz, and P. Grassberger. Performance of different synchronization measures in real data: A case study on electroencephalographic signals. *Phys. Rev. E*, 65(041903):1–14, 2002.

[46] N. Marwan and J. Kurths. Nonlinear analysis of bivariate data with cross recurrence plots. *Phys. Lett. A*, (302):299–307, 2002.

[47] F. Censi, V. Barbaro, P. Bartolini, G. Calcagnini, and S. Cerutti. Non–linear dynamics of atrial rate during atrial fibrillation assessed by recurrence plot analysis. *IEEE Computers in Cardiology Proc.*, 25:197–200, September 1998.

[48] N. Marwan, N. Wessel, U. Meyerfeldt, A. Schirdewan, and J. Kurths. Recurrence–plot–based measures of complexity and their application to heart–rate–variability data. *Phys. Rev. E*, 66(026702):1–8, 2002.

[49] D. B. Geselowitz. On the theory of the electrocardiogram. *Proc. of the IEEE*, 77(6):857–876, June 1989.

[50] J. D. Z. Chen, B. D. Schirmer, and R. McCallum. Measurements of electrical activity of the human small intestine using surface electrodes. *Trans. on Biomed. Eng.*, 40(6):598–602, June 1993.

[51] I. Taylor, H. L. Duthie, R. Smallwood, and D. Linkens. Large bowel myoelectrical activity in man. *Gut*, (16):808–814, July 1975.

[52] C. J. Stoddard, H. L. Duthie, R. H. Smallwood, and D. A. Linkens. Colonic myoelectrical activity in man: comparison of recording techniques and methods of analysis. *Gut*, (20):476–483, 1976.

[53] M. A. Allessie, A. P. G. Hoeks, G. M. L. Schmitz, and R. S. Reneman. On–line mapping system for the visualization of the electrical activation of hte heart. *Int. J. Cardiol. Imag.*, (2):59–63, 1986.

[54] W. J. E. P. Lammers, A. Al-Kais, S. Singh, K. Arafat, and T. Y. El-Sharkawy. Multi–electrode mapping of slow wave activity in the isolated rabbit duodenum. *J. Appl. Physiol.*, (74):1454–1461, 1993.

[55] W. J. E. P. Lammers, B. Stephen, K. Arafat, and G. W. Manfield. High resolution electrical mapping in the gastrointestinal system: initial results. *Neurogastrointerol. Mot.*, (8):207–216, 1996.

[56] W. J. E. P. Lammers. Propagation of individual spikes as patches of activation in isolated feline duodenum. *Am. J. Physiol. Gastrointest. Liver Physiol.*, (278):G297–G307, 2000.

[57] B. O. Familoni, Y. J. Kingma, and K. L. Bowes. Noninvasive assessement of human gastric motor function. *IEEE Trans. on Biomed. Eng.*, BME–34(1):30–36, January 1987.

[58] B. O. Familoni, T. L. Abell, and K. L. Bowes. A model of gastric electrical activity in healt and disease. *IEEE Trans. on Biomed. Eng.*, 42(7):647–657, July 1995.

[59] M. Mintchev and K. Bowes. Computer simulation of the effect of changing abdominal thickness on the electrogastrogram. *Med. Eng. and Phys.* 20:177–18l, 1998.

[60] L. A. Bradshaw, W. O. Richards, and J. P. Wikswo. Volume conductor effects on the spatial resolution of magnetic fields and electric potentials from gastrointestinal electrical activity. *Med. Biol. Eng. Comput.*, 39:35–43, 2001.

[61] W. J. E. P. Lammers and J. R. Slack. Of slow waves and spike patches. *News Physiol. Sci.*, 16:138–144, June 2001.

[62] P. Z. Rashev, M. P. Mintchev, and K. L. Bowes. Application of an object–oriented programming paradigm in three–dimensional computer modeling of mechanically active gastrointestinal tissues. *IEEE Trans. Inform. Technol. Biomed.*, 4(3):247–258, September 2000.

[63] P. Z. Rashev K. L. Bowes, and M. P. Mintchev. Three–dimensional object–oriented modeling of the stomach for the purpose of microprocessor–controlled functional stimulation. *IEEE Trans. Inf. Tech. Biomed.*, 6(4):296–309, December 2002.

[64] P. Z. Rashev, M. P. Mintchev, and K. L. Bowes. Three–dimensional static parametric modeling of phasic colonic contractions for the purpose of microprocessor–controlled functional stimulation. *J. Med. Eng. Tech.*, 25(3):85–96, June 2001.

[65] P. Z. Rashev, M. Amaris, K. L. Bowes, and M. P. Mintchev. Microprocessor–controlled colonic peristalsis. *Dig. Dis. Sci.*, 47(5):1034–1048, May 2002.

[66] R. N. Miftakhov and G. R. Abusheva. Numerical simulation of excitation-contraction coupling in a locus of the small bowel. *Biol. Cybern.*, (74):455–467, 1996.

[67] R. N. Miftakhov, G. R. Abusheva, and J. Christensen. Numerical simulation of motility patterns of the small bowel. 1. formulation of a mathematical model. *J. theor. Biol.*, (197):89–112, 1999.

[68] N. J. Spencer and T. K. Smith. Simulataneous intracellular recordings from longitudinal and circular muscle during the peristaltic reflex in guinea–pig distal colon. *J. Physiol.*, 533(3):787–799, 2001.

[69] T.-Y. Lee, P.-H. Lin, C.-H. Lin, Y.-N. Sun, and X.-Z. Lin. Interactive 3–d virtual colonscopy system. *IEEE Trans. on Information Technology in Biomedicine*, 3(2):139–150, June 1999.

[70] L. Hong, Z. Liang, A. Viswambharan, A. Kaufman, and M. Wax. Reconstruction and visualization of 3d models of colonic surface. *IEEE Trans. on Nucl. Sci.*, 44(3):1297–1302, June 1997.

[71] S. K. Sarna, E. E. Daniel, and Y. J. Kingma. Simulation of slow–wave electrical activity of small intestine. *Am. J. Physiol.*, 221(1):166–175, July 1971.

[72] G. V. Osipov and M. M. Sushchik. The effect of natural frequency distribution on cluster synchronization in oscillator arrays. *Trans. on Circ. and Syst.*, 44(10):1006–1010, 1997.

[73] V. Schlageter, P.-A. Besse, R. S. Popovic, and P. Kučera. Tracking system with five degrees of freedom using a 2d–array of hall sensors and a permanent magnet. *Sensors and Actuators A*, 92:37–42, 2001.

[74] V. Schlageter, P. Drljaca, R. S. Popovic, and P. Kučera. A magnetic tracking system based on highly sensitive integrated hall sensors. *JSME Int. J. Series C*, 45:967–973, 2002.

[75] V. Schlageter. *Magnetic tracking system for clinical investigation of gastrointestinal motility*. PhD thesis, Lausanne Swiss Federal Institute of Technology, 2003.

[76] S. Nishida, M. Nakamura, A. Ikeda, and H. Shibasaki. Signal separation of background eeg and spike by using morphological filter. *Med. Eng. and Phys.*, 21:601–608, 1999.

[77] L. Vincent. Morphological trasformations of binary images with arbitrary structuring elements. *Sig. Proc.*, 22:3–23, 1991.

[78] R. Jones and I. Svalbe. Algorithms for the decomposition of gray–scale morphological operations. *IEEE Trans. Pattern Anal. Machine Intell.*, 16(6):581–588, June 1994.

[79] F. Y.-C. Shih and O. R. Mitchell. Threshold decomposition of gray–scale morphology into binary morphology. *IEEE Trans. Pattern Anal. Machine Intell.*, 11(1):31–42, January 1989.

[80] R. L. Stevenson and G. R. Arce. Morphological filters: statistics and further syntactic properties. *IEEE Trans. on Circ. Sys.*, CAS–34(11):1292–1305, November 1987.

[81] W. A. Voderholzer, W. Schatke, B. E. Muhldorfer, A. G. Klauser, B. Birkner, and S. A. Muller-Lissner. Clinical response to dietary fiber treatment of chronic constipation. *Am. J. Gastroenterol.*, 92:95–98, 1997.

[82] D. C. Nyam, J. H. Pemberton, D. M. Ilstrup, and D. M. Rath. Long–term results of surgery for chronic constipation. *Dis. Colon Rectum*, 40:273–279, 1997.

[83] W. J. Snape, S. A. Matarazzo, and S. Cohen. Effect of eating and gastrointestinal hormones on human colonic myoelectrical and motor activity. *Gastroenterol.*, 75(3):373–378, 1978.

[84] I. Taylor, H. L. Duthie, R. Smallwood, B. H. Brown, and D. Linkens. The effect of stimulation on the myoelectrical activity of the rectosigmoid in man. *Gut*, 15:599–607, 1974.

[85] A. G. Klauser, J. Flaschentrager, A. Gehrke, and S. A. Muller-Lissner. Abdominal wall massage: effect on colonic function in healthy volunteers and in patients with chronic constipation. *Gastroenterol.*, 30(4):247–251, April 1992.

[86] T. Hirabayashi, H. Matsufuji, J. Yokoyama, K. Hagane, K. Hoshino, Y. Morikawa, and M. Kitajima. Colorectal motility induction by sacral nerve electrostimulation in a canine model. *Dis. Colon Rectum*, 46(6):809–817–, June 2003.

[87] A. C. Malouf, P. H. Wiesel, T. Nicholls, R. J. Nicholls, M. Chir, and M. A. Kamm. Short–term effects of sacral nerve stimulation for idiopathic slow transit constipation. *World J. Surg.*, 26(2):166–170, February 2002.

[88] B. O. Familoni, T. L. Abell, D. Nemoto, G. Voeller, and B. Johnson. Efficacy of electrical stimulation at a frequency higher than basal rate in canine stomach. *Dig. Dis. Sci.*, 42(5):892–897, May 1997.

[89] B. O. Familoni, T. L. Abell, G. Voeller, A. Salem, and O. Gaber. Electrical stimulation at a frequency higher than basal rate in human stomach. *Dig. Dis. Sci.*, 42(5):885–891, May 1997.

[90] R. W. McCallum, J. D. Z. Chen, Z. Lin, B. D. Schirmer, R. D. Williams, and R. A. Ross. Gastric pacing improves emptying and symptoms in patients with gastroparesis. *Gastroenterology*, 114:456–461, March 1998.

[91] M. P. Mintchev, C. P. Sanmiguel, M. Amaris, and K. L. Bowes. Microprocessor–controlled movement of solid gastric content using sequential neural electrical stimulation. *Gastroenterology*, 118:258–263, 2000.

[92] X. Lin, J. Hayes, L. J. Peters, and J. D. Z. Chen. Entrainment of intestinal slow waves with electrical stimulation using intraluminal electrodes. *Ann. Biomed. Eng.*, 28:582–587, 2000.

[93] C. A. Mosse, T. N. Mills, M. N. Appleyard, S. S. Kadirkamanathan and C. P. Swain. Electrical stimulation for propelling endoscopes. *Gastrointest. Endosc.*, 54(1):79–83, 2001.

[94] S. F. Hughes, S. M. Scott, M.-A. Pilot, and N. S. Williams. Electrically stimulated colonic reservoir for total anorectal reconstruction. *Br. J. Surg.*, 82(10):1321–1326, October 1995.

[95] A. Maw, M.-A. Pilot, M. R. Hutton, A. J. P. Eccersley, A. Basu, M. Scott, and N. S. Williams. The effect of direct electrical stimulation on the intact colon. *Br. J. Surg.*, 85:14, 1998.

[96] K. Bruninga, L. Riedy, A. Keshavarzian, and J. Walter. The effect of electrical stimulation on colonic transit following spinal cord injury in cats. *Spinal Cord*, 36:847–853, 1998.

[97] M. A. Amaris, P. Z. Rashev, M. P. Mintchev, and K. L. Bowes. Microprocessor controlled movement of solid colonic content using sequential neural electrical stimulation. *Gut*, 50:475–479, 2002.

[98] A. Shafik, O. El-Sibai, and A. A. Shafik. Rectal pacing in patients with constipation due to rectal inertia: technique and results. *Int. J. Colorectal. Dis.*, 15:100–104, 2000.

[99] A. Shafik and O. El-Sibai. Rectal pacing: pacing parameters required for rectal evacuation of normal and constipated subjects. *J. of Surg. Res.*, 88(2):258–263, February 2000.

[100] A. Shafik, A. A. Shafik, O. El-Sibai, and I. Ahmed. Colonic pacing in patients with constipation due to colonic inertia. *Med. Sci. Monit.*, 9:CR191–CR196, 2003.

[101] A. Moritz, S. Grundfest-Broniatowski, L. Ilyes, J. Kasick, G. Jacobs, E. Olsen, and Y. Nose. Electrical pulse train and single pulse stimulation of the small intestine: acute and chronic studies in the dog. *Artif Organs.*, 13(2):116–122, April 1989.

[102] S. Aellen, P. H. Wiesel, J.-P. Gardaz, V. Schlageter, M. Bertschi, N. Virag, and J.-C. Givel. Electrical stimulation induces propagated colonic contractions in an experimental model. *Br. J. Surg.*, 96(2):214–220, 2009.

[103] P. Berčik, D. Armstrong, R. Fraser, P. Dutoit, C. Emde, M.-P. Primi, A.-L. Blum, and P. Kučera. Origins of motility patterns in isolated arterially perfused rat intestine. *Gastroenterol.*, 106(3):649–657, March 1994.

[104] K. Hipper and H. J. Ehrlein. Motility of the large intestine and flow of ingesta in pigs. *Res. in Vet. Sci.*, 71:93–100, 2001.

[105] W. J. E. P. Lammers, S. Dhanasekaran, J. R. Slack, and B. Stephen. Two–dimensional high–resolution motility mapping in the isolated feline duodenum: methodology and initial results. *Neurogastroenterol. Mot.*, 13:309–323, 2001.

[106] A. Moritz, S. Grundfest-Broniatowski, L. Ilyes, J. Kasick, G. Jacobs, and Y. Nose. Contractile response to electrical stimulation of the small intestine in anesthetized and awake dogs. *Artif. Organs.*, 13(6):553–557, December 1989.

[107] A. M. Accarino, F. Azpiroz, and J. R. Malagelada. Symptomatic responses to stimulation of sensory pathways in the jejunum. *Am. J. Physiol.*, 263(5.1):G673–G677, November 1992.

[108] A. M. Accarino, F. Azpiroz, and J-R. Malagelada. Gut perception in humans is modulated by interacting gut stimuli. *Am. J. Physiol. Gastrointest. Liver. Physiol.*, 282:G220–G225, 2002.

[109] J. L. Martinez de Juan, J. Saiz, M. Meseguer, and J. L. Ponce. Small bowel motility: relationship between smooth muscle contraction and electroenterogram signal. *Med. Eng. and Phys.*, 22:189–199, 2000.

Die VDM Verlagsservicegesellschaft sucht für wissenschaftliche Verlage abgeschlossene und herausragende

Dissertationen, Habilitationen, Diplomarbeiten, Master Theses, Magisterarbeiten usw.

für die kostenlose Publikation als Fachbuch.

Sie verfügen über eine Arbeit, die hohen inhaltlichen und formalen Ansprüchen genügt, und haben Interesse an einer honorarvergüteten Publikation?

Dann senden Sie bitte erste Informationen über sich und Ihre Arbeit per Email an *info@vdm-vsg.de*.

Sie erhalten kurzfristig unser Feedback!

VDM Verlagsservicegesellschaft mbH
Dudweiler Landstr. 99
D - 66123 Saarbrücken
www.vdm-vsg.de

Telefon +49 681 3720 174
Fax   +49 681 3720 1749

Die VDM Verlagsservicegesellschaft mbH vertritt

Printed by Books on Demand GmbH, Norderstedt / Germany